Jerry Stachowski
8/23/75

LOUDSPEAKERS AND LOUDSPEAKER CABINETS

# LOUDSPEAKERS AND LOUDSPEAKER CABINETS

P. W. VAN DER WAL

**Drake Publishers Inc.**
**381 Park Avenue South**
**New York, N.Y. 10016**

Published 1970

**Drake Publishers Inc**
**381 Park Avenue South**
**New York, N.Y. 10016**

SBN 87749-002-3

## FOREWORD

As interest in sound-reproduction has increased, so also has the demand for loudspeaker boxes enabling the response characteristics of the speaker to be exploited to the full. What is often forgotten, however, is that other factors affecting response have to be taken into account, as well as the box. For example, the acoustics of the room are just as important as the matching of the speaker to the amplifier.

Besides providing particulars of numerous different speaker boxes, this book is intended to be a source of information on various matters which, although apparently incidental, must nevertheless be given due consideration if the best results are to be obtained. To enable the "do-it-yourself" enthusiast to deal with certain problems which may arise during the construction of the boxes, a chapter of practical hints is also included.

January 1966 P. W. VAN DER WAL

## CONTENTS

CHAPTER 1

# THE LOUDSPEAKER

## Purpose

The task of the loudspeaker is to convert alternating electric current into the physical movement of a cone. This movement sets up vibrations in the air, which are, in turn, converted by our sense of hearing into sound within the 16 to 16,000 c/s frequency range. We cannot actually hear vibrations of higher frequency than 16,000 c/s, but, if they are strong enough, we may feel them as pain. Frequencies below 16 c/s may even produce feelings of mental discomfort or fear.

## How it works

The basic principle of the loudspeaker is the ability of a vibrating diaphragm to produce audible vibrations. In practice, such vibrations are set up by means of the force exerted on a conductor in a magnetic field. The magnitude of this force depends on the current passing through the conductor, the length of the latter and the intensity of the magnetic field. This relationship is linear in the case of all three of the parameters.

In the formula

$$F = B \times i \times l$$

(i.e. force = magnetic field intensity × current × length of wire), the MKSA scale equates the concept of force ($F$), the intensity of the magnetic field or magnetic inductance ($B$), the current ($i$) and length ($l$) with the Newton ($Vs/m^2$), Ampere and metre units. The direction of the force, usually called the "Lorentz force", is determined by the direction of the current through the conductor and the direction of the magnetic field.

The direction of the force may be deduced from the directions of the current and the flux by means of the "corkscrew rule" (see fig. 1), in that it

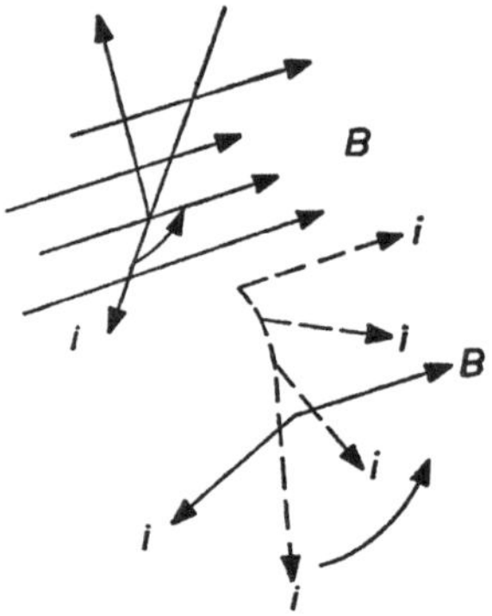


Fig. 1. Conductor in a magnetic field. The direction of the force acting on the conductor may be found with the aid of the corkscrew rule.

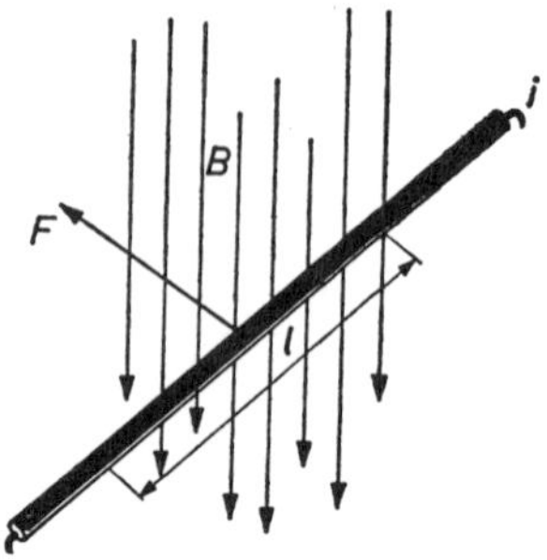


Fig. 2. Conductor of length $l$ in a magnetic field.

corresponds to the direction of thrust of the "corkscrew" rotated in such a way that the direction of the current, rotated through the smallest angle coincides with that of the field.

Fig. 2 shows a conductor with a current $i$ (amps) flowing through it. The conductor is in a field of magnetic inductance $B$ (Vs/m2). The force exerted on the portion of the wire of length $l$ (m) is $B \times i \times l$ (newtons), whilst its direction may be deduced from the corkscrew rule. In point of fact, the formula $F = B \times i \times l$ is accurate only when the wire crosses the direction of the field at right angles. A correction factor must be applied when the angle of intersection is less than 90°, and this factor depends on the size of the angle, although it will always be less than unity. Formulated:

$$F = k \times B \times i \times l,$$

where $k$ is the correction factor equal to cos $\psi$, $\psi$ being the angle between the directions of flux and current.

So much for the very simple magnetic field in fig. 2. Let us now take a look at the conductor in the field shown in fig. 3. Here, the wire has been bent into a circle and positioned in the magnetic field in such a way as to be concentric with it. As before, the wire and lines of force cross at right angles, so that the formula $F = B \times i \times l$ again applies. The length of wire that can be introduced into a field of given size can be very much increased by winding it into a coil. This also affects the force exerted on the wire, since $F$ increases with $l$ in the formula $F = B \times i \times l$. This method is used, say, in the construction

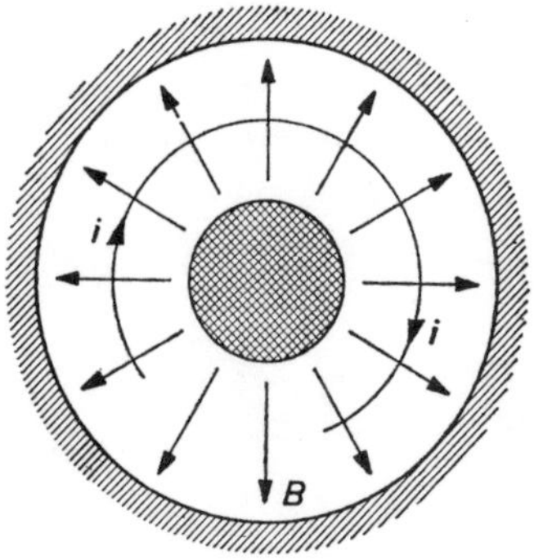


Fig. 3. Radial magnetic field with circular conductor.

Fig. 4. Speaker coil.

of loudspeakers. Wire is wound around a tube of impregnated paper to form the speaker coil (fig. 4), which moves in the magnetic field of the speaker. This field may be set up by a system of magnets (see fig. 5) comprising a permanent magnet (1), two pole-pieces (2 and 3) and a soft-iron core (4). In this way, a uniform magnetic field of high intensity, having the flux pattern shown in fig. 6, is produced in the air-gap (5).

When a.c. currents are passed through the speaker coil, the amplitude and frequency variations of the coil will both correspond to the fluctuations in the current. A heavy current will produce major deviations due to the magnitude of $F$, then equal to $B \times i \times l$. A change in the direction of the current (a.c.) will produce a change in the direction of the force exerted on

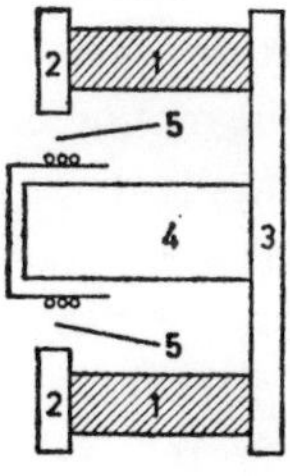


Fig. 5. System of magnets in a speaker.

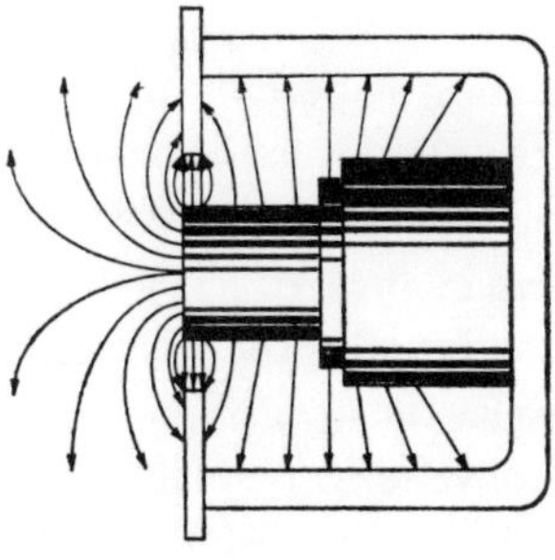

Fig. 6. Magnetic field in the air-gap, with lines of force.

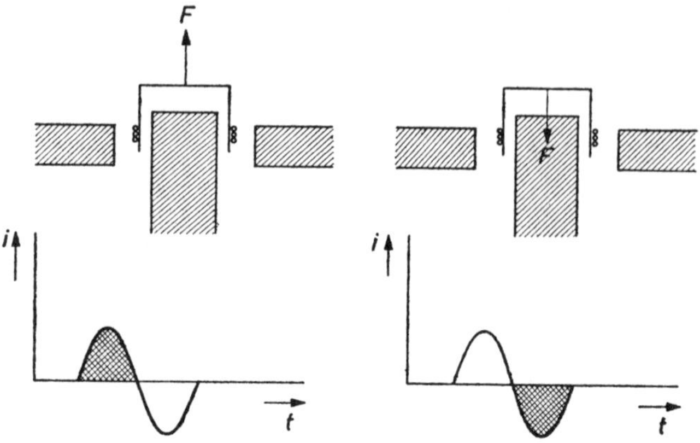


Fig. 7. Diagrammatic representation of the cone movements with the current flowing in different directions.

the speaker coil, thereby reversing the movements of the latter, as shown diagrammatically in fig. 7.

Although we have now shown how the speaker coil is caused to move, we have still not explained how this movement is made to produce sound. To do this, the speaker cone (1 in fig. 8), which may be made of special paper, is secured to the speaker coil. To allow the cone to imitate the movements of the coil, its edge is crimped (2). The cone is centred by the centring ring (3) and suspended in the cone support (4). The result is that any movements of the speaker coil are repeated by the cone, which, in turn, sets up air vibrations. Our ears can now interpret this vibration of the air in terms of sound.

Because it is perhaps a little surprising that the speaker cone is usually, albeit not always, made of paper, it is worthwhile taking a closer look at the stresses set up in the speaker. What happens is that the motion of the cone is governed by the frequency in that this determines not only the rhythm of the vibration but also the vibrating area of the cone. The higher the frequency, the smaller the area of the cone that actually vibrates, the reason being that the effective mass is smaller at high than at low frequencies.

Now the cone must accommodate itself to such effective mass-reduction, otherwise the response to stimuli received from the speaker coil at high frequencies will be inadequate owing to the inertia of the system constituted by

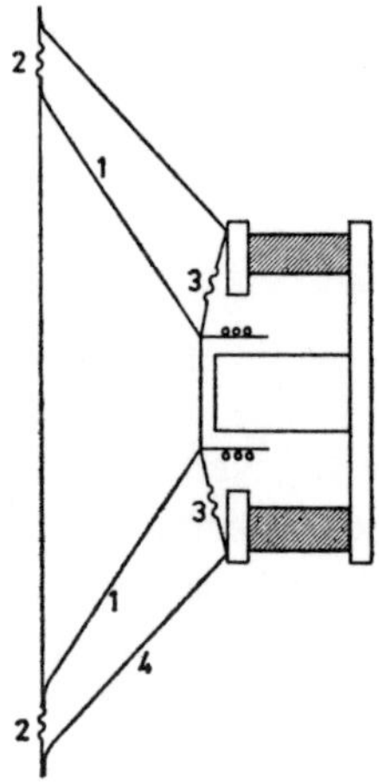


Fig. 8. The speaker.

the speaker coil and the complete cone, the coil being unable to change its effective mass. Paper generally, and the specially impregnated cone-paper in particular, permits such effective mass-reduction (fig. 9; facing p. 8).

At the same time, much more rigid cones are sometimes used for special bass loudspeakers. Because such cones cannot accommodate to mass-reduction, the bass speaker does not respond to high frequencies.

Loudspeakers are of many different types, which may be easily distinguished by examining the technical data.

## Technical data

### *Power handling capacity* (watts)

This is the maximum electrical power that can be applied to the speaker without damaging it (fig. 10a), equivalent to $E \times i \times \cos \psi$, or, in words, voltage across speaker coil times current through coil times the cosine of the phase angle between current and voltage.

### *Speaker coil impedance* ($\Omega$)

By this is meant the a.c. resistance of the speaker, usually measured at

1,000 c/s because of the frequency-dependence (fig. 10b). The coil impedance of low-impedance loudspeakers is 2, 4, 8, 10 and 15 Ω and that of the high-impedance types is 400 or 800 Ω.

*Efficiency (%)*

The efficiency of the speaker is the percentage ratio between the acoustic and electric power produced by and applied to it (fig. 10c).

Because it is governed by the frequency, the efficiency is measured at a standard test frequency of 400 c/s. It is determined by the intensity of the magnetic field and the design and characteristics of the speaker cone, air-gap and speaker coil. The space factor is particularly important in this connection. It is the quotient of the volumes of speaker coil and air-gap. Efficiency increases with the space factor.

*Resonant frequency (c/s)*

The cone vibrates most readily at what is known as its resonant, or natural, frequency (fig. 10a). This is easily recognisable as a peak in the frequency

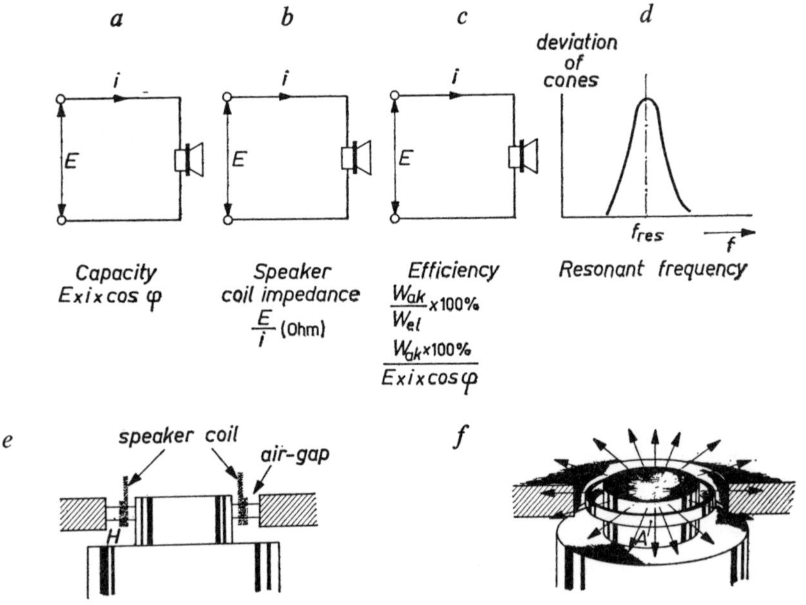


Fig. 10. Diagrammatic representation of speaker values.

response curve. It is also the lowest frequency to which the speaker responds; towards lower frequencies the curve falls off 12 dB per octave.

As will be explained more fully later, the task of the speaker box is to ensure that low frequencies reproduced by the actual speaker duly reach our ears instead of being suppressed by "acoustic short-circuiting". Some boxes are prone to this, thereby producing what is, to all intents and purposes, a rise in the resonant frequency.

*Magnetic field intensity (A/m, formerly Oersted)*

$$1 \text{ Oersted} = \frac{4\pi}{10^3} \text{ A/m}$$

The magnetic field intensity ($H$) is associated with the magnetic flux density $B$ through the equation $B = \mu H$. Here, $\mu = \mu_0 \times \mu_r$, where $\mu_0$ (absolute permeability) is a constant and $\mu_r$ (relative permeability) is a factor depending on the material (fig. 10e); $\mu_r$ is unity in air and very much greater than unity in ferromagnetic materials. $\mu_0 = 4\pi \times 10^{-7}$ Vs/A.

The efficiency of the speaker is proportional to the field intensity and therefore depends very much upon it.

*Magnetic flux (Wb = Vs, formerly Maxwell)*

$$1 \text{ maxwell} = 10^8 \text{ weber.}$$

The magnetic flux in the air-gap is the product of the area $A$ of the speaker coil surface (fig. 10f) around the air-gap and the component of the magnetic inductance (or flux density) at right angles to this surface. Formulated: $Ø = B \times A$, or, in words, flux = màgnetic inductance × area. The efficiency of the speaker is likewise proportional to the magnetic flux.

*Magnet weight (kg)*

This is the weight of the permanent magnet, expressed in kg in the MKSA scale. It appears that the term magnet weight is often misinterpreted. The idea that a strong magnetic field (in the air-gap) can be obtained only with the aid of a heavy magnet is false. The fact is that the optimum weight of a magnet also depends very largely on the quality of its steel. For example, it may well happen that a magnet weighing 1.0 kg is used to produce a magnetic field that could equally well be supplied by a 0.1 kg type of a better quality steel. Ticonal magnetic steel having a *BH* product of as much as $6.4 \times 10^4$ Wb A/m$^3$ is used in Philips speakers.

*Speaker weight (kg)*

This is the gross weight of the loudspeaker.

*Speaker dimensions*

The type of speaker varies according to its dimensions, standardised by RETMA*) specifications; hence, the choice of speaker depends very much on the space available for the speaker box in the particular installation. The principal speaker dimensions, often specified with an accompanying diagram in technical publications, are the overall depth, the pitch circle of the fixing holes and the baffle aperture. The baffle aperture is the hole made in that wall of the box behind which the speaker is fixed (fig. 11).

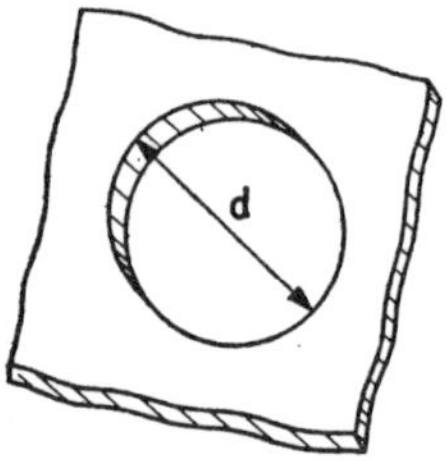


Fig. 11. Baffle with aperture.

At the same time there is another factor in that undesirable side-effects may result from the use of timber more than 20 mm thick, owing to resonance of the air in the cylindrical "chamber" formed by the hole in front of the speaker. This cylindrical column of air acts as a Helmholtz resonator, with all its annoying results (fig. 12). If, nevertheless, relatively thick timber has to be used for any reason, as may happen in the case of acoustic boxes and bass-reflex cabinets (chapter 7), it is best to recess and bevel off the portion of the baffle nearest the speaker (fig. 13). Some dimensions of loudspeakers are given in fig. 14.

Let us pause at this point to consider one or two details of the construction. Our analysis of the cone movements was based on the assumption that they

*) RETMA: Radio Electronic & Television Manufacturers' Association.

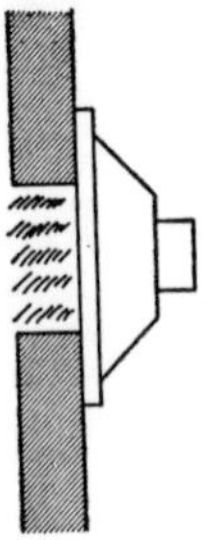

Fig. 12. Speaker with deep baffle aperture; Helmholtz resonator.

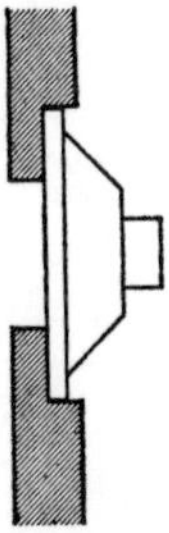

Fig. 13. Speaker with reduced-depth baffle aperture.

are governed solely by the frequency and the fluctuation of the current in the speaker coil. This is valid only provided that the flux enclosed by the coil remains constant. It may happen, however, that the movement of the coil produces sudden changes in the amount of flux embraced by it, thereby changing the force exerted on the coil in a manner not dictated by the current

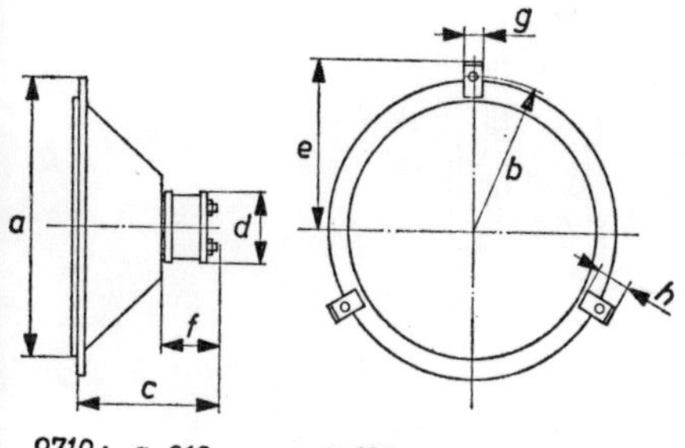


9710: a= 216 b= 113 c= 116 d= 74 e= 128 f= 60 g= 15 h= 33

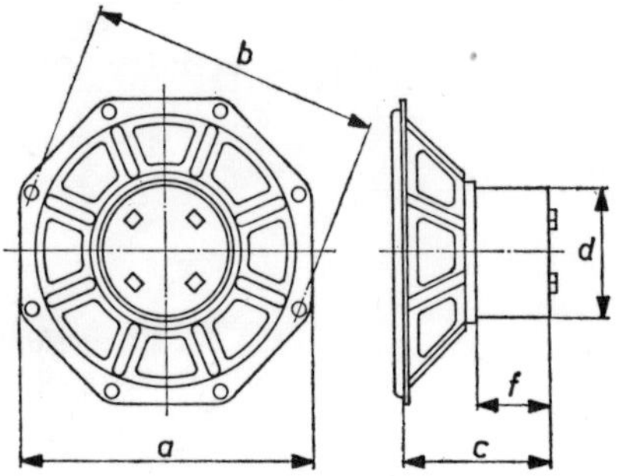


AD 4800: a=191,6 b=194 c=120,2 d= 76 f= 59,2

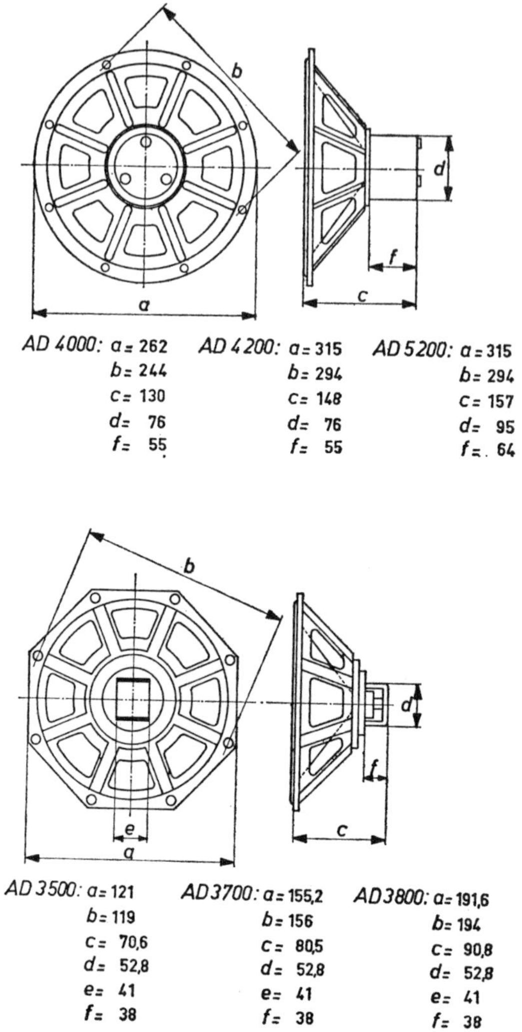


Fig. 14. Dimensions of various speakers.

(fig. 15). Because the current in the speaker coil is in turn related to and, in the ideal case, identical with (albeit on a smaller scale) the current fed to the input of the amplifier by, say, a pick-up or tape recorder, any sudden change of flux may produce a sound pattern other than that originally recorded on

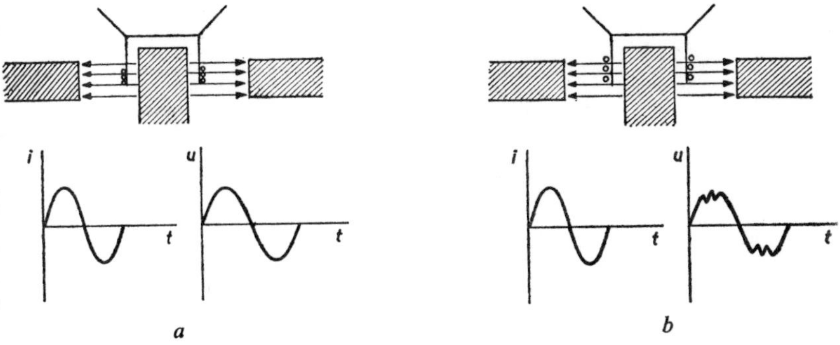


Fig. 15. Diagrammatic representation of cone movements produced by an a.c. current flowing through the speaker coil. *a.* deflections take place as dictated, *b.* not exactly so (distortion) because of the fact that the coil occasionally strays from the magnetic field

the particular tape or gramophone record. This is what is known as distortion (chapter 4). To eliminate this form of distortion from Philips Hi-Fi loudspeakers, the ratio of the height of the air-gap to that of the speaker coil is so calculated as to enable the coil to absorb the effect of any undesired flux variations. This can be done by making the size of the air-gap greater than the height of the speaker coil (fig. 16).

Fig. 16. Speaker with air-gap larger than the height of the coil.

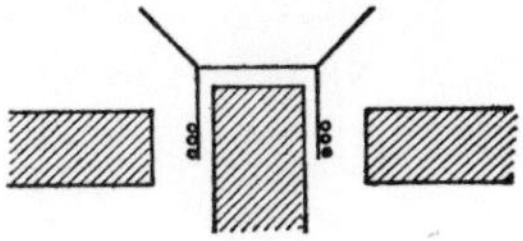

One of the more important design features from the point of view of the efficiency is the width of the air-gap. A wide gap is the simplest, since it permits wider tolerances and makes the shape of the speaker coil less critical. At the same time, a wide air-gap lowers the magnetic field intensity, thereby necessitating larger magnets and better-quality steel. By far the best solution, and one which also improves the uniformity of the magnetic field, is to find the best combination of high-quality steel and air-gap size. For example, fig. 17 shows the speaker coil dimensions and air-gap size of the Philips AD 3700 speaker. It will be evident from this that close tolerances make great demands on production methods and particularly on the centring of the speaker coil.

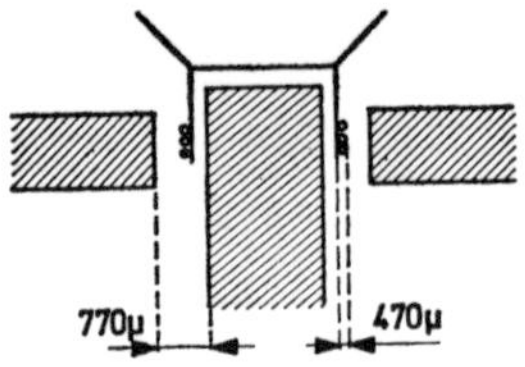


Fig. 17. Dimensions of coil and air-gap of Philips AD 3700 speaker.

Before going on to discuss some of the graphs, which are most important for the assessment of a loudspeaker, let us examine the technical data of certain speakers (page 13) whose boxes are illustrated in chapter 9.

In addition to the aforementioned data, technical specifications often contain diagrams providing information on the performance of the speaker at different frequencies. These graphs are entitled sound-pressure curve and radiation pattern.

## Sound-pressure curve

The sound-pressure curve shows the relationship between the sound-pressure produced by the loudspeaker and the frequency of the a.c. voltage delivered to the speaker coil. On the MKSA scale, the sound-pressure is measured in newtons per square metre ($N/m^2$). Fig. 18 gives an example of a sound-pressure curve, with the frequency plotted on the horizontal and the sound-pressure $N/m^2$ on the vertical axis.

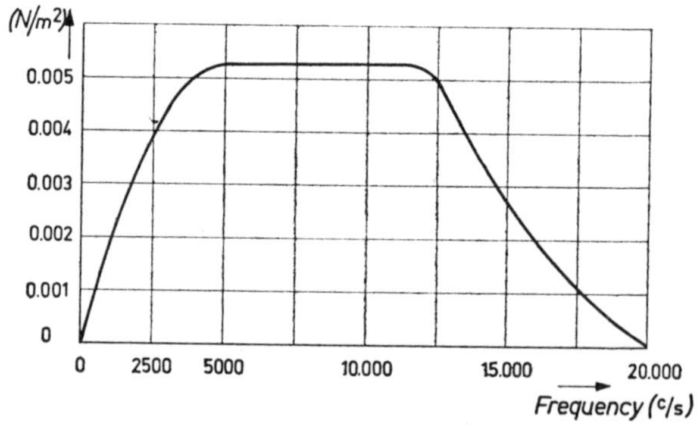


Fig. 18. Theoretically correct but practically useless sound-pressure curve.

| Loudspeaker type | Max. loadability | Voice coil imp. (1000 c/s) | Efficiency (400 c/s) | Resonant frequency | Flux density | | Magnetic flux | | Magnet weight | Total weight | Baffle aperture |
|---|---|---|---|---|---|---|---|---|---|---|---|
| | Watt | Ohm | % | c/s | $Vs^1 / m^2$ | Gauss | Vs (Wb) | Maxwell | kg | kg | m |
| AD 3500/06 | 3 | 5 | 4 | 130 | 1.1 | 11,000 | $262 \cdot 10^{-6}$ | 26,200 | 0.077 | 0.400 | 0.105 |
| AD 3500/06 M | 3 | 5 | 4 | 130 | 1.1 | 11,000 | $262 \cdot 10^{-6}$ | 26,200 | 0.077 | 0.400 | 0.105 |
| AD 3500/06 AM | 3 | 800 | 4 | 130 | 1.1 | 11,000 | $262 \cdot 10^{-6}$ | 26,200 | 0.077 | 0.400 | 0.105 |
| AD 3700/06 | 3 | 5 | 6 | 90 | 1.1 | 11,000 | $262 \cdot 10^{-6}$ | 26,200 | 0.077 | 0.440 | 0.141 |
| AD 3700/06 M | 3 | 5 | 6 | 90 | 1.1 | 11,000 | $262 \cdot 10^{-6}$ | 26,200 | 0.077 | 0.440 | 0.141 |
| AD 3700/06 AM | 3 | 800 | 6 | 90 | 1.1 | 11,000 | $262 \cdot 10^{-6}$ | 26,200 | 0.077 | 0.440 | 0.141 |
| AD 3800/06 | 6 | 5 | 6 | 75 | 1,1 | 11,000 | $262 \cdot 10^{-6}$ | 26,200 | 0.077 | 0.520 | 0.176 |
| AD 3800/06 M | 6 | 5 | 6 | 75 | 1.1 | 11,000 | $262 \cdot 10^{-6}$ | 26,200 | 0.077 | 0.520 | 0.176 |
| AD 3800/06 AM | 6 | 800 | 6 | 75 | 1.1 | 11,000 | $262 \cdot 10^{-6}$ | 26,200 | 0.077 | 0.520 | 0.176 |
| AD 4000 | 10 | 7 | 6 | 50 | 0.8 | 8,000 | $98 \cdot 10^{-5}$ | 98,000 | 0.298 | 1.770 | 0.227 |
| AD 4000 M | 10 | 7 | 6 | 50 | 0.8 | 8,000 | $98 \cdot 10^{-5}$ | 98,000 | 0.298 | 1.770 | 0.227 |
| AD 4000 AM | 10 | 800 | 6 | 50 | 0.8 | 8,000 | $98 \cdot 10^{-5}$ | 98,000 | 0.298 | 1,770 | 0.227 |
| AD 4200 | 20 | 7 | 7 | 45 | 0.8 | 8,000 | $98 \cdot 10^{-5}$ | 98,000 | 0.298 | 1.800 | 0.276 |
| AD 4200 M | 20 | 7 | 7 | 45 | 0.8 | 8,000 | $98 \cdot 10^{-5}$ | 98,000 | 0.298 | 1.800 | 0.276 |
| AD 4200 AM | 20 | 800 | 7 | 45 | 0.8 | 8,000 | $98 \cdot 10^{-5}$ | 98,000 | 0.298 | 1.800 | 0.276 |
| AD 4800 | 6 | 5 | 10 | 60 | 1.3 | 13,000 | $583 \cdot 10^{-6}$ | 58,300 | 0.375 | 1.500 | 0.176 |
| AD 4800 M | 6 | 5 | 10 | 60 | 1.3 | 13,000 | $583 \cdot 10^{-6}$ | 58,300 | 0.375 | 1.500 | 0.176 |
| AD 5200 | 20 | 7 | 14 | 45 | 1.1 | 11,000 | $134 \cdot 10^{-5}$ | 134,000 | 0.830 | 3.500 | 0.276 |
| AD 5200 M | 20 | 7 | 14 | 45 | 1.1 | 11,000 | $134 \cdot 10^{-5}$ | 134,000 | 0.830 | 3.500 | 0.276 |
| AD 5200 AM | 20 | 800 | 14 | 45 | 1.1 | 11,000 | $134 \cdot 10^{-5}$ | 134,000 | 0.830 | 3.500 | 0.276 |
| 9710 | 10 | 7 | 5 | 50 | 0.8 | 8,000 | $97 \cdot 10^{-5}$ | 97,000 | 0.428 | 1.800 | 0.195 |
| 9710 M | 10 | 7 | 5 | 50 | 0.8 | 8,000 | $97 \cdot 10^{-5}$ | 97,000 | 0.428 | 1.800 | 0.195 |
| 9710 AM | 10 | 800 | 5 | 50 | 0.8 | 8,000 | $97 \cdot 10^{-5}$ | 97,000 | 0.428 | 1,800 | 0.195 |

**M = Double cone**

At the same time, it is more usual to graduate the coordinates in a manner different from that shown in fig. 18. The graduation is linear from 0 to 100 c/s ("normal") along the axis, but becomes "logarithmic" beyond this. The logarithmic system gives a very compact, and therefore convenient scale, covering the wide range of frequencies from 100 to 20,000 c/s and closely matched to the audible range. The reason for the linear scale from 0 to 100 c/s is that these frequencies are particularly important and could not be depicted clearly enough logarithmically.

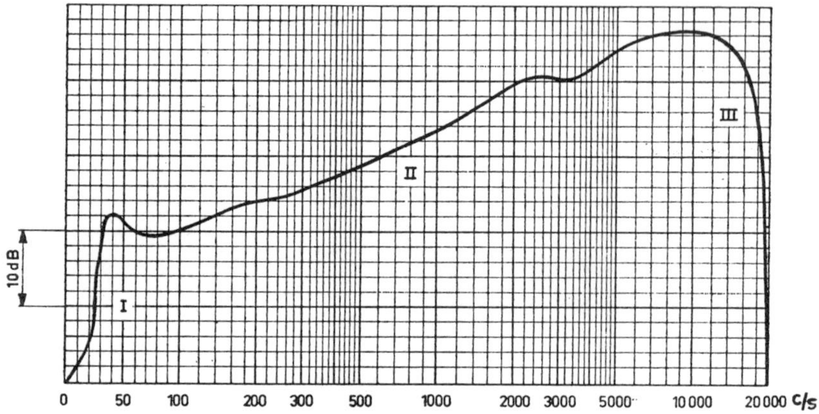


Fig. 19. Example of a sound-pressure curve.

Fig. 19 shows a sound-pressure curve of the kind found in publications. The most notable feature is that the vertical and horizontal axes have different scales, apparently linear, but actually complex.

Another important factor in sound-reproduction is the relative amplitude of the different sounds. In theory, it would be possible, and indeed logical, to transform the sound-pressure curve in fig. 18 into another as shown in fig. 20, in which the sound-pressure at 1000 c/s is taken as a unit to express the sound-pressures at other frequencies. That this graph is, in fact, of little practical value is because of a quirk in our sense of hearing which makes the ratio of two amplitudes $A_1$ and $A_2$ at the same frequency sound more like the logarithm of $A_1/A_2$ than $A_1/A_2$ itself (fig. 21).

This logarithm of sound-intensities is called the bel; however, the unit generally used is the decibel (dB), one-tenth of a bel. For example, when $\log A_1/A_2 = a$, the ratio of $A_1$ to $A_2$ is expressed as: $a$ bel, or $10a$ dB.

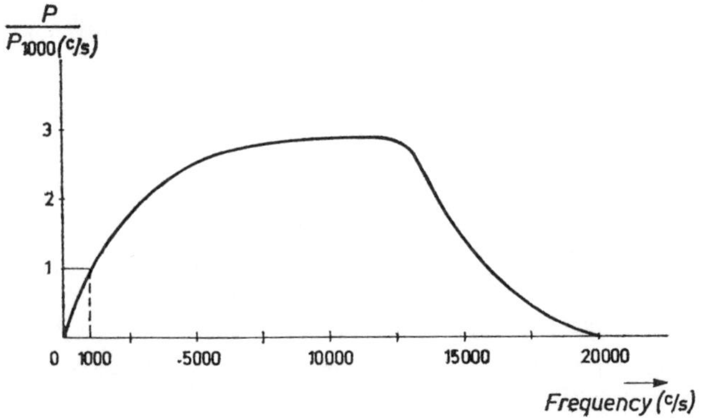


Fig. 20. The same curve as in fig. 18 but with better divisions on the coordinates.

In electroacoustics, the dB is also used to express power ratios. Thus $a$ dB $= 10 \log W_1/W_2$, where $W_1$ and $W_2$ are the wattages involved. In a comparison of sound-pressure, voltage and current, however, $b$ dB $= 20 \log P_1/P_2$ or $20 \log E_1/E_2$ or $20 \log I_1/I_2$, the factor of 20 being due to the fact that these values have to be squared to produce powers on which to base the calculation in dB's.

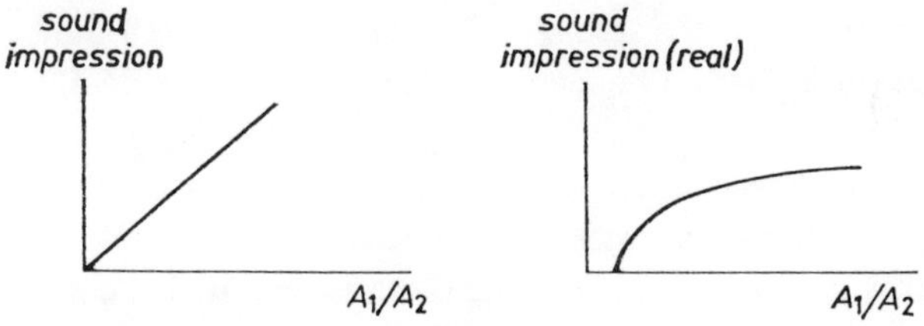


Fig. 21. Graphic representation of the amplitude ratios as heard by the human ear if it reacts linearly (*a*) and as it really does perceive them (*b*).

In graphic representations in which some value, say, the sound-pressure in dB's is plotted along one of the axes, a zero dB level must be established. As compared with a sound-pressure of 1.5 $N/m^2$, 3.0 $N/m^2$ is greater by $20 \log \frac{3.0}{1.5} = 6$ dB, whereas 3.0 $N/m^2$ referred to 1 $N/m^2$ gives $20 \log \frac{3.0}{1.0} = 9.5$ dB. The reference level, or "zero dB level", is therefore very important, since without it, the graph is meaningless. For example, with a sound-pressure of 2.0 $N/m^2$ as the reference level, a measured sound-pressure of 2.0$N/m^2$ will be plotted on the dB axis as 0 dB, since $20 \log \frac{2.0}{2.0} = 20 \log 1 =$ 0 dB. The sound-pressure curves in this book have a zero dB level at $2 \times 10^{-5}$ $N/m^2$. The smallest audible change in amplitude is 1 dB.

So much for the units and the way in which they are set out along the axes; let us now take another look at the sound-pressure curve shown in fig. 19. The resonant frequency is readily recognisable as a characteristic peak in the curve, occurring in the region of 40 c/s in fig. 19. Note that the characteristic is divided into three sections, I, II and III. Section I provides information concerning the response of the speaker in the range below resonance. In this section, the sensitivity of the speaker exhibits a sharp drop of at least 12 dB per octave throughout the range concerned. Without the baffle, it may even diminish by as much as 18 dB per octave.

With a view to the matters to be discussed in Chapter 7, it is necessary to stress the importance of differentiating between the sound-pressure curves of speakers with and without a baffle, since the two respond entirely differently to bass frequencies, generally below 1000 c/s. The sound-pressure curve of a loudspeaker without a baffle is not nearly as good as that of one mounted in a box or on a baffle. Plotted without a box, the speaker characteristic shows a drop of 18 dB below the resonant frequency, whilst with the baffle effect extending towards lower frequencies (Chapter 7), a drop of 12 dB will be observed.

For a good loudspeaker, section II of the curve rises steadily, since, above 1000 to 2000 c/s, the sound-waves reproduced become more closely concentrated towards higher frequencies. The result is that, with a constant energy output, the sound-pressure immediately in front of the speaker will undergo a steady increase.

Section III tends to fall off again. At the same time, it does not follow that the transition from section II to section III represents the highest frequency to which the speaker can respond. This is taken to be the frequency at which the sound-pressure is the same as at 1000 c/s, that is 19,500 c/s in the case illustrated in fig. 19.

Following this explanation of the sound-pressure curve in theory, a few words on the practical method of plotting it will not be out of place.

A calibrated microphone is placed 50 cm from the front of the speaker in a "dead" room completely lined with sound-absorbent material (fig. 22; facing p. 8). A signal of constant amplitude whose frequency fluctuates between 0 and 20,000 c/s is delivered to the speaker, and the voltage put out by the microphone is a measure of the sound-pressure. The deflection of the stylus of an automatic recorder is governed by this voltage.

Some features of the speaker may be demonstrated with the aid of the sound-pressure curve, as follows. A double-cone speaker (fig. 23) is used to obtain optimum treble response. Up to 10,000 c/s it is the large cone which undergoes effective mass-reduction, whilst above this frequency the smaller one takes over. Fig. 24 shows two sound-pressure curves for the Philips

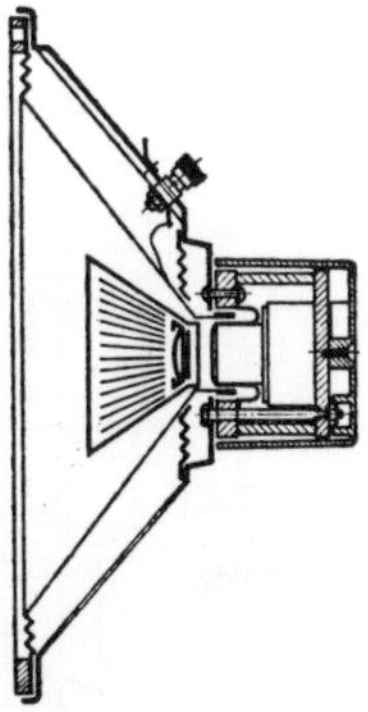

Fig. 23. A double-cone speaker.

loudspeaker AD 3800, with and without a second cone. The curves are identical up to 3000 c/s, but show a marked divergence beyond this.

As we have seen in passing, the impedance of the speaker coil depends upon the frequency. Since the impedance of the speaker coil has to be matched to the internal resistance of the amplifier (see chapter 3), such dependence is not conducive to good response. Moreover, it can be kept within reasonable bounds by screening the magnetic core with copper (see fig. 25).

In connection with the method of measuring the sound-pressure curve, we mentioned a signal of constant amplitude applied to the loudspeaker. The rather loose term "amplitude" may be applied equally to voltage, current or energy. With constant current through the speaker coil, the voltage across it fluctuates with the frequency. Conversely, the current varies in the same way

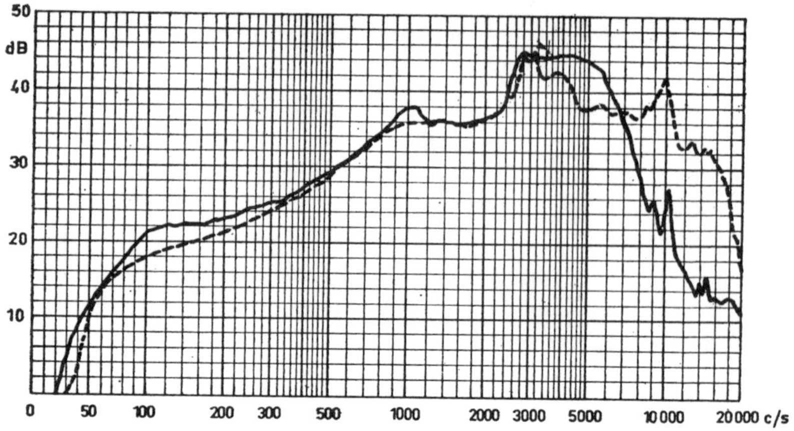


Fig. 24. Sound-pressure curves of Philips AD 3800 speaker with a special double cone.

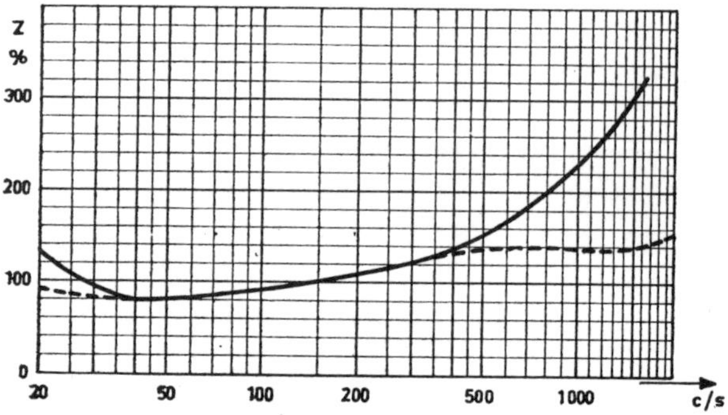


Fig. 25. Uncorrected (curve *a*) and corrected (curve *b*) coil impedance curves.

when the voltage is constant. Either constant-voltage or constant-current curves are obtained, depending on whether it is the voltage across or the current in the coil that is kept constant. These two types of curve are shown in fig. 26. The constant-current curve has a very sharp peak at the resonant frequency, whilst the constant-voltage curve shows lower sound-pressures towards higher frequencies. Thus the former is a better guide for the loud-speaker designer, whereas the latter is of more direct use to the user, since most amplifiers employ feedback and are, therefore, in effect, constant-voltage generators. The sound-pressure curves of the speakers described and

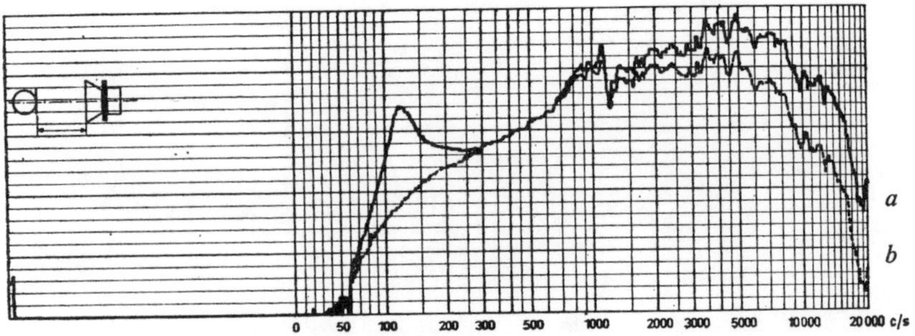


Fig. 26. Sound-pressure curves at constant current (*a*) and voltage (*b*).

recommended in this book are given on page 104 ff. The shape of these curves shows that they were plotted without a baffle and at constant voltage

## The radiation pattern

The radiation pattern likewise defines a relationship between the sound-pressure and the frequency. Unlike the sound-pressure curve, however, which is plotted with the microphone on the axis of the speaker, the radiation pattern is established by applying a constant frequency to the voice coil and varying the angle between the speaker axis and the line from speaker to microphone (fig. 27). The ratio of sound-pressure to angle is measured at several frequencies, the results being shown in fig. 28. Accordingly, the

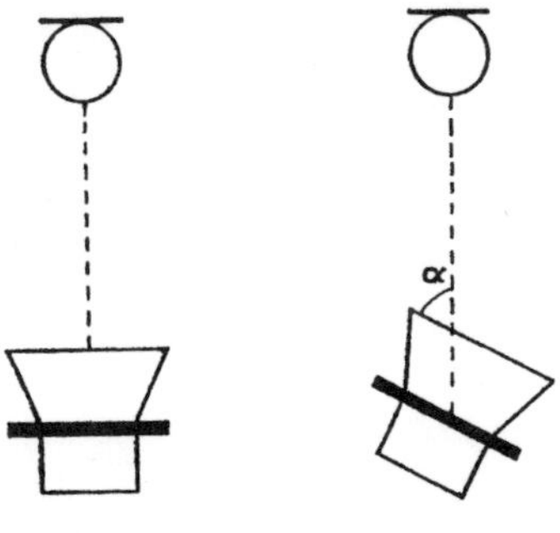


Fig. 27. Measuring the radiation pattern.

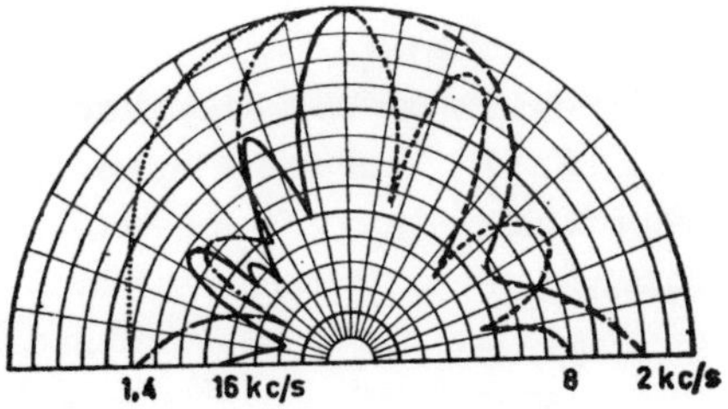


Fig. 28. Radiation pattern of Philips AD 4800 M speaker.

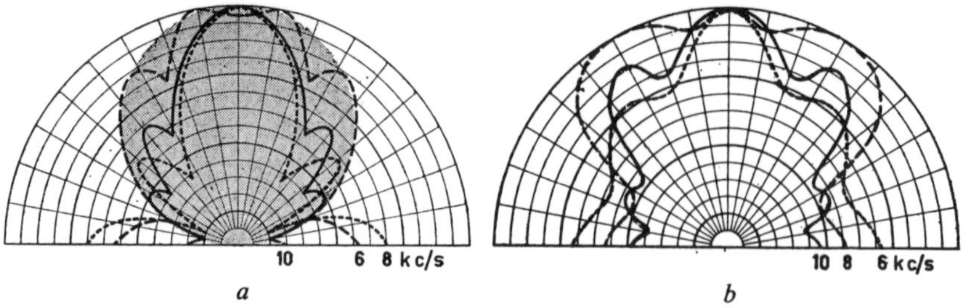


Fig. 29. Radiation pattern of Philips 9710 M speaker; single-cone model (*a*) and double-cone model (*b*).

radiation pattern illustrates the diffuse rendering of sound by a speaker. For instance, fig. 29 shows the radiation pattern of a double-cone loudspeaker together with that of a single-cone type.

So far, only the electrodynamic speaker has been considered. The electrostatic and crystal types will now be discussed, and we shall also be taking a brief look at two newly developed systems used in the iso- and orthophase speakers.

Electrostatic speakers are used mainly as tweeters. They operate on the principle that two parallel metal plates either attract or repel each other according to their charge (fig. 30).

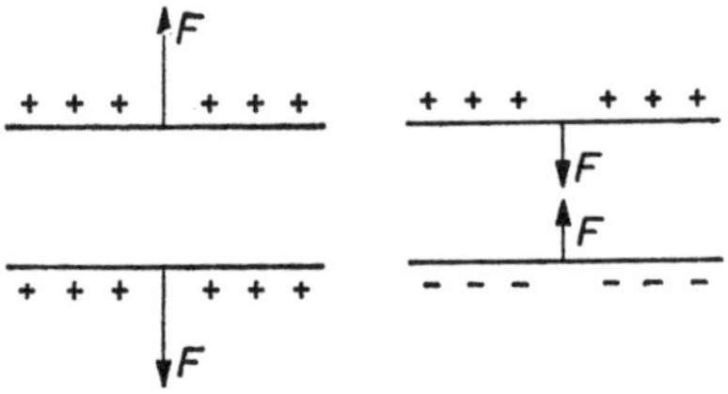


Fig. 30. Principle of the electrostatic speaker.

Low voltages can be produced by the deformation of a crystal (the piezo-electric effect), one of whose uses is in the crystal pick-up head. The opposite effect is used in the crystal loudspeaker, where reproduction is restricted to frequencies above 5000 c/s, since the crystals tend to shatter at low frequencies. This type of speaker is not often used owing to its very irregular sound-pressure curve.

The vibrating surface of the isophase speaker is formed by a fibrous synthetic material with a conductor applied in a sinuous pattern to one or

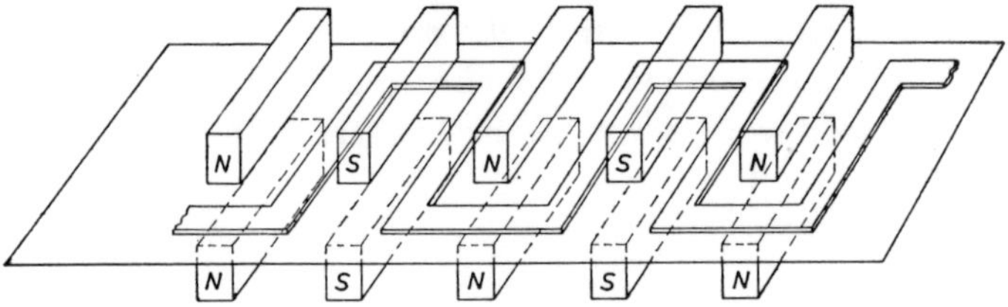


Fig. 31. Diagram of isophase speaker.

both sides (fig. 31). Magnets are arranged on opposite sides of the material in such a way as to place the conductor in the strongest possible magnetic field (figs. 32 and 33). This type of speaker has very good treble response up to almost 40,000 c/s, whilst its bass response is determined by the area of the material and the voltage applied to it. Apart from physically enlarging the area of the material in one speaker, the effective area can also be increased by connecting several isophase loudspeakers in parallel or series. Apertures are made in the "walls" of the speaker through which the air can be forced out (fig. 34; facing p. 9).

Fig. 32. Location of the magnet in relation to the conductor in an isophase speaker.

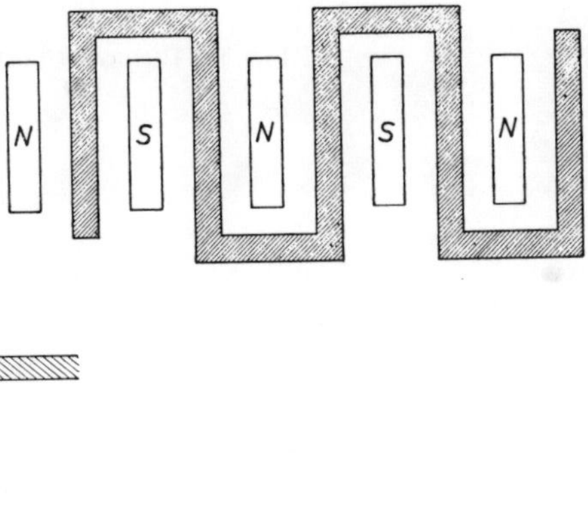


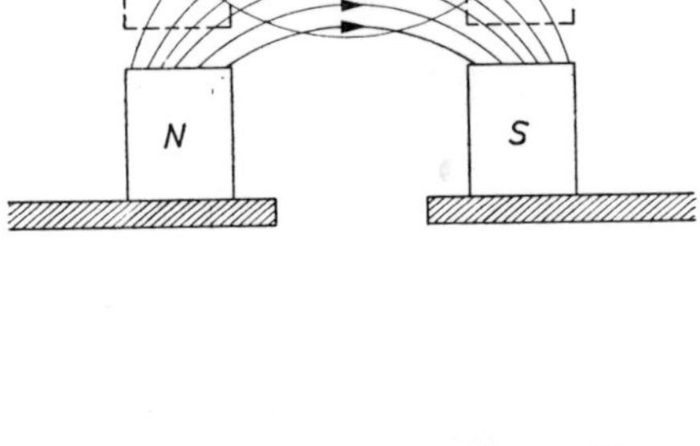


Fig. 33. Section through conductor in a magnetic field.

There is one more type, namely the orthophase loudspeaker, the construction of which is shown in fig. 35. Here, too, a fibrous synthetic material provided with conductors is used, but magnets are arranged on one side only.

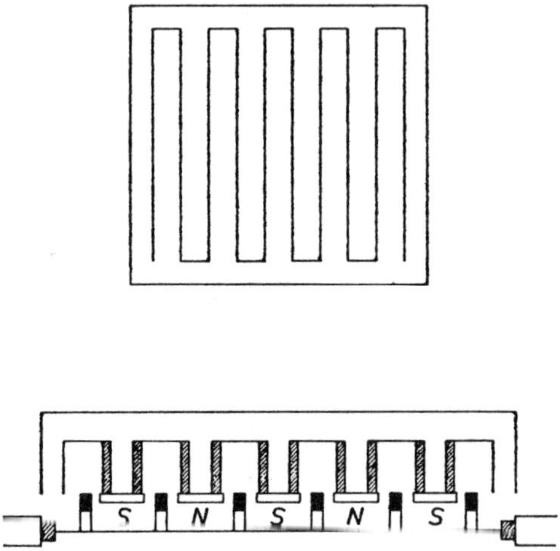


Fig. 35. Diagram of orthophase speaker.

CHAPTER 2

# CHOOSING A LOUDSPEAKER

A number of factors have to be considered in choosing the right type of speaker for a particular installation, and these factors, in their turn, are determined by the purpose envisaged for the installation and the design of the amplifier. Apart from these, there are also such important factors as the output wattage, impedance and the distortion.

The load imposed on the speaker must be matched to the output wattage of the amplifier. Obviously, an amplifier capable of supplying 20 W, for instance, cannot be connected to a speaker capable of taking 6 W, since the latter will inevitably be damaged by greater wattages than 6 W. A certain amount of distortion is produced with the amplifier at maximum power (see chapter 4), e.g. at 20 W, $D = 2\%$, so that a 20 W speaker must be used in this case. Fig. 36 gives the amplifier power required to reproduce the loudest (fff) and quieter "forte" (ff and f) sounds in rooms of various sizes.

Apart from loadability, speaker coil impedance is also important, and distinction should be made between high-impedance (400 or 800 Ω) and low-impedance (4 to 20 Ω) speakers. Again, different impedances are used for special applications, transistorised circuitry, for instance. The speaker coil impedance must be matched to the output impedance of the amplifier, so that a low-impedance amplifier with an output impedance of 10 Ω must be connected to a speaker with speaker coil impedance also of 10 Ω. If the impedance of a low-impedance speaker is twice or three times this value, i.e. at least 20 to 30 Ω, two of these speakers must be used in parallel, and a transformer may also be employed (see chapter 3).

There is obviously no point in using a poor amplifier with a very high-quality speaker. If, on the other hand, the amplifier is a good one, i.e. produces very little distortion and gives adequate amplification, it is spoiling the ship to use a loudspeaker giving a fair amount of distortion.

It is possible to determine the quality of a speaker from its sound-pressure curve and radiation pattern. Both a low resonant frequency, of the order of

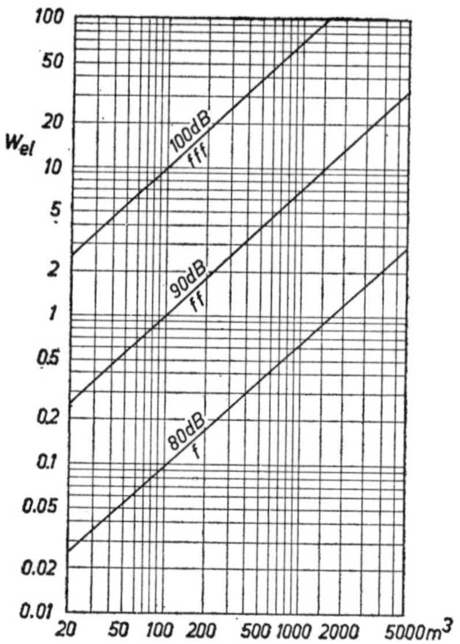


Fig. 36. Graph of relationship between the volume of the room, output power and sound ntensity, with a speaker efficiency of 5%.

50 c/s. and a faithful treble response are necessary for quality sound-reproduction, and these requirements may be satisfied by the use of a double-cone speaker. The effect of a "constant" speaker coil impedance is essential to really high-quality reproduction. In this case, too, however, there is still little point in using a very mediocre amplifier in conjunction with a speaker having this very pleasant quality. By and large, the speaker efficiency must be as high as possible, although efficiency may have to be sacrificed, in order to obtain the best response, by extra attenuation. A good example of this is the Philips 9710 speaker, in which the cone and the magnet are specially shaped to provide a cushion of air between them (fig. 37), thus giving a very fine curve (fig. 38), with an efficiency of 5%.

It is not, of course, quite so essential to use a double-cone speaker where filters are used to limit response to only a part of the frequency spectrum and a speaker with a good bass response is chosen. The double-cone type is best used as a tweeter, since bass response is immaterial in this case (fig. 39).

An important factor in the choice of the speaker box is the size of the

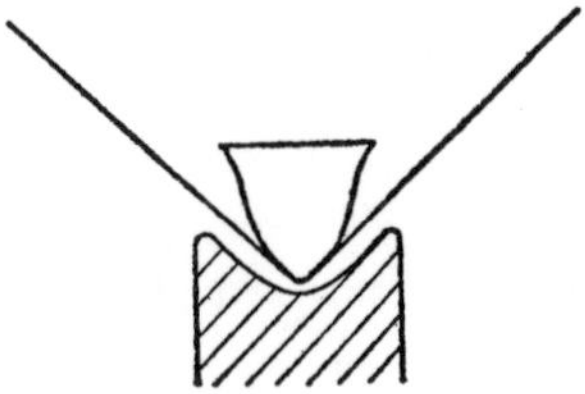

Fig. 37. Type of damping used in Philips 9710 M speaker.

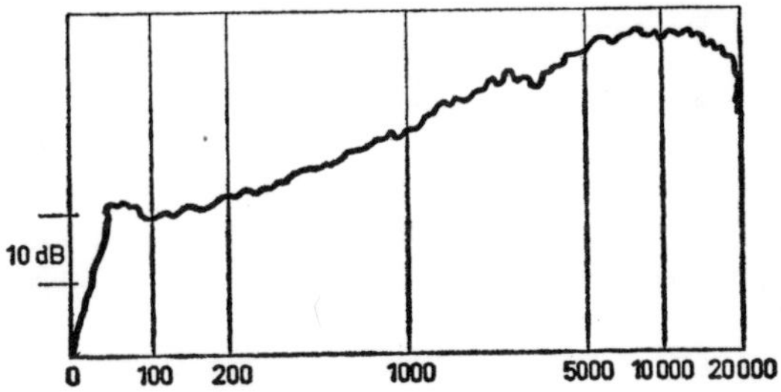


Fig. 38. Sound-pressure curve of Philips 9710 M speaker.

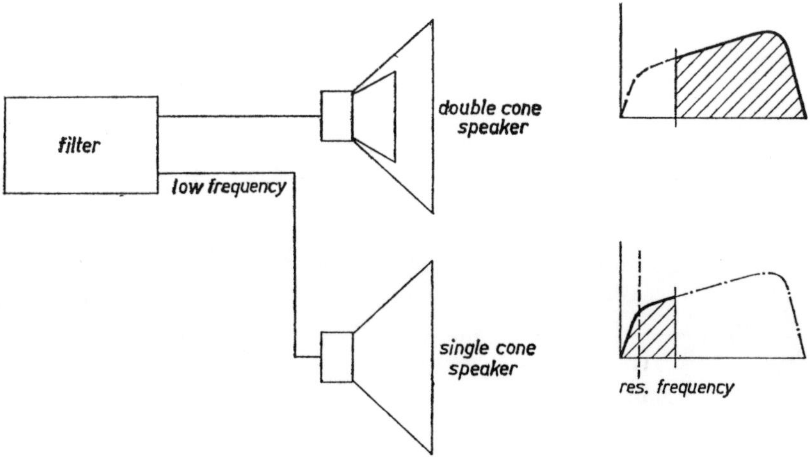


Fig. 39. Diagrams of types of speaker used in combination with a filter. The sound-pressure curves are cross-hatched in the zone important to response.

speaker itself (chapters 9 and 10). The significant factors are given in the summary below:

1. Load capacity, and matching to the wattage of the amplifier.
2. Speaker coil impedance may be high or low, depending on the amplifier. If the output and speaker coil impedance differ fairly widely, a transformer or parallel or series circuit may be used.

3. Resonant frequency of about 50 c/s for good bass response.
4. A double-cone speaker for the higher frequencies (above 10,000 c/s).
5. Efficiency, amplifier power and distortion are very closely interrelated. In the case of the 9710, a special attenuator design gives very good response, but with smaller speakers it generally results in lowered efficiency.
6. Pick the most suitable speakers for use with a filter.

CHAPTER 3

# A FEW FUNDAMENTALS

This chapter will be dealing with some of the most essential points in a sound-reproduction installation, which, if ignored, will inevitably lead to the eventual quality's suffering, and may well lead to a faulty assessment of the components used in the installation.

## The weakest link

An important point, but one that is often overlooked, is that a chain (of sound, for instance), is no stronger than its weakest link. In other words, a poor amplifier can never give good results, however good the rest of the components in the layout. This applies, of course, not merely to the amplifier, but to all the other components as well.

## Matching the speaker to the amplifier

The job of the amplifier is to deliver as much electrical power as possible to the speaker, which converts this power into sound-waves. The more power the loudspeaker receives, the more sound it will reproduce. It must always be remembered, however, that this power is limited by the loadability of the loudspeaker itself.

The latter is now connected to the amplifier (fig. 40). If they are properly matched, the speaker will receive the maximum power, whereas poor matching may seriously affect the power-transfer. This will now be explained with the aid of a simplified analogy comprising a voltage-source of internal resistance $R_i$ connected to a load resistance $R_b$ (fig. 41). The source and load respectively represent the amplifier and speaker. The current $i$ flowing in the circuit is $\frac{E}{R_i + R_b}$ Ampere, while the power dissipated by the resistance is

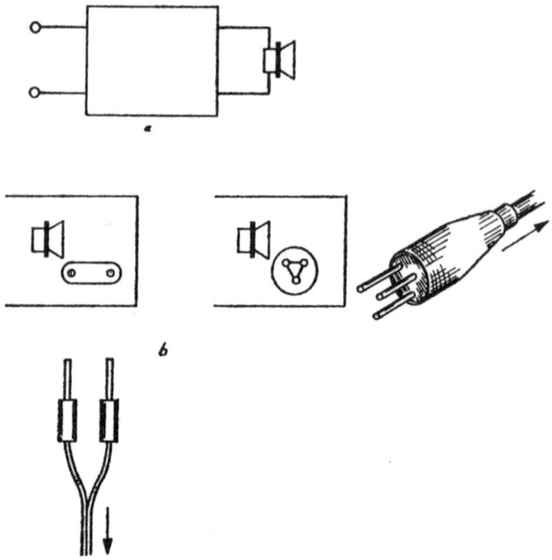

Fig. 40. Diagram showing the connection of the speaker to the amplifier, with the connection for a low-impedance speaker on the left and that for a high-impedance type on the right. "Banana plugs" are used for the former, whilst, for safety reasons, only special plugs may be used for the latter.

$$R_b \frac{E^2}{(R_i + R_b)^2} \text{ Watts.}$$

This power is at its peak when $R_i = R_b$ and, assuming that the power taken by the speaker from the amplifier is required to be as high as possible, the speaker coil impedance ($R_b$ in this example) must be equal to the internal resistance of the amplifier ($R_i$ in this example). If this is, in fact, so, then the speaker and amplifier are properly matched.

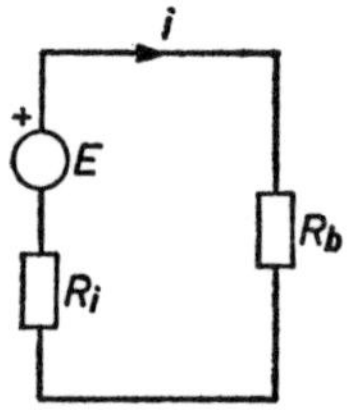


Fig. 41. Idealised circuit diagram of the connection of speaker to amplifier.

If now, this example is taken even further, and it is assumed that the $R_i$ of the voltage-source is 800 Ω, and a value of 400 Ω has been chosen for impedance $R_b$, the power supplied to the resistance will be only 40% of what it could be with an $R_b$ of 800 Ω. This makes the importance of proper matching abundantly clear.

In the two examples that have just been used we have made two simplifications, which have no effect on the ultimate result but serve to illustrate the point. The load resistance has been regarded as a straight resistance, although it is being used to represent the speaker coil impedance, which, in point of fact, comprises a resistance and a self-inductance. This is also true of the internal resistance. Another simplification is that the more complex action of the anode impedance on the final-stage valve of the amplifier has been disregarded. By considering $R_i$ and $R_b$ as constants this reaction has been ignored. The impedance to which an amplifier has to be connected is called the matching impedance, or, less correctly, matching resistance.

These notes show how to provide an amplifier with a loudspeaker of suitable impedance, although the number of possibilities open to us is restricted by the fact that, with an existing amplifier, a speaker has to be found having both acoustic properties, etc, and an impedance suitable for this amplifier. Where, however, both amplifier and speaker are already in our possession and are not properly matched, a matching transformator will have to be used.

Basically, a transformer consists of two windings, each made up of a different number of turns, but wound on the same core, generally of soft iron (fig. 42). If an a.c. voltage is applied to the primary winding, i.e. the transformer input, the resultant current will produce a magnetic field of alter-

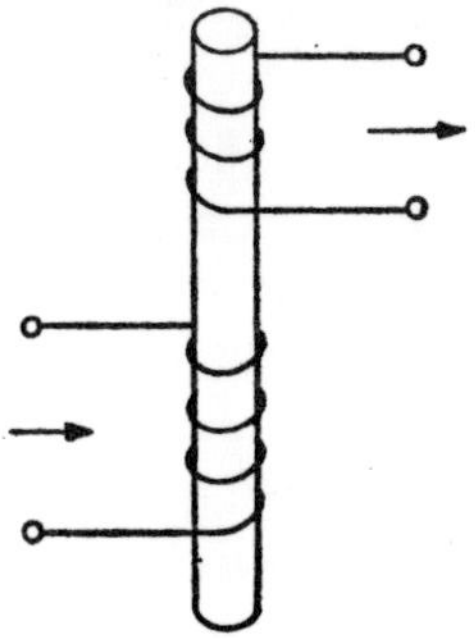

Fig. 42. Simple transformer

nating direction and intensity, whose lines of force will pass more readily through the soft iron than through air, owing to the resistance of the latter, and therefore the magnetic field will be highly concentrated in the core. Since the secondary winding is also wound around it, any variation in the magnetic field will excite voltages in this winding, and, if an impedance is connected to it, a current will flow. The intensity of the magnetic field is determined by the current and the number of turns in the primary winding, which, together with the winding ratio and the size of the voltage applied to the transformer input, also determines the voltage across the secondary winding. The winding ratio of a transformer is the ratio between the number of turns in the primary and secondary windings. Fig. 43 shows a diagram of a transformer, showing the soft iron core and the winding ratio, and fig. 44 shows a transformer connected to an impedance (e.g. a speaker) of $Z$ Ω. If the winding ratio is $n : 1$, the input impedance of a transformer connected to impedance $Z$ is $n^2Z$ Ω. Hence a speaker of impedance $Z$ may be connected to an amplifier of impedance $n^2Z$ Ω, so that, where the speaker impedance is 5 Ω, it may be connected to a 180 Ω amplifier via a transformer with six primary turns and one secondary.

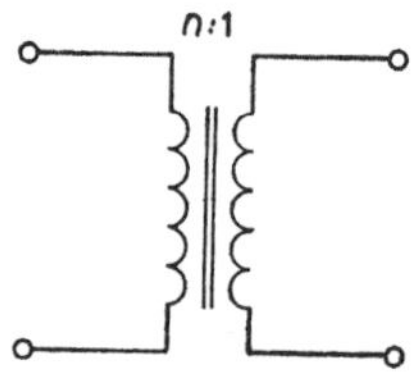


Fig. 43. Diagram of a transformer.

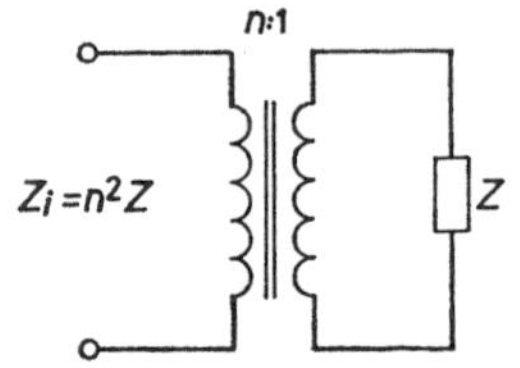


Fig. 44. Transformer connected to impedance $Z$.

Having dealt with the matching transformer, let us now look at high- and low-impedance loudspeakers, illustrated in figs. 45 and 46 respectively. The transformer in fig. 45 is used as a matching transformer to match the low-impedance speaker to the several-thousand Ω output impedance of the amplifier. Fig. 46, however, shows no transformer, since here the speaker coil impedance is already high enough not to need one, and requires only a special circuit in which $Z$ may be between 500 and 1000 Ω instead of the usual 6000 Ω. The drawback to a matching transformer is the distortion produced and the power-loss that it involves.

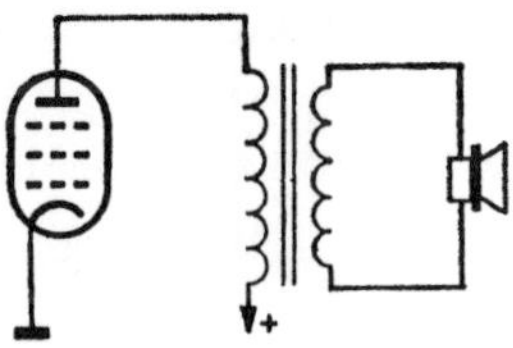

Fig. 45. Final stage valve of amplifier with output transformer and loudspeaker (low-impedance output).

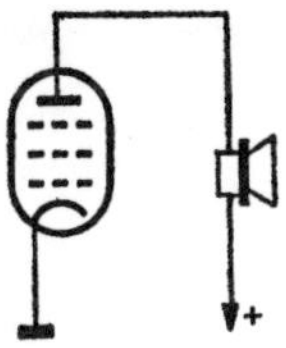

Fig. 46. Speaker directly connected to valve (high-impedance output).

## High-impedance amplifiers

Now a word about high-impedance amplifiers to round off the above remarks. If, for any reason, the connections between speakers and high-impedance amplifiers have to be interrupted, it is best to cut out the amplifier. If the amplifier is not switched off, but the speaker is cut off instead, the position shown in fig. 47 obtains, where the final-stage valve has to absorb all the power, and this may lead to its destruction. The same is true, although to a lesser extent, of low-impedance amplifiers.

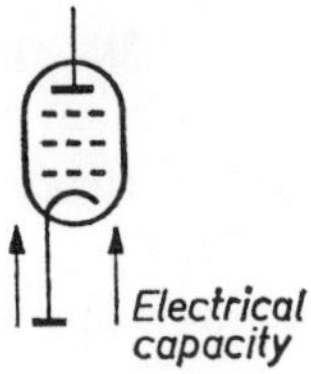


Fig. 47. If the speaker connection breaks down while the amplifier is in operation, all the output power will have to be dissipated by the final-stage valve.

## The connection of speakers in parallel and in series

Where a number of loudspeakers is connected to a monaural amplifier or the outputs of a stereo amplifier, to obtain a stereophonic effect, for example (see chapter 5), they may be connected in series or parallel. The manner of connection depends upon the impedances of the amplifier and speaker and the wattage involved. In theory, four speakers of 10 Ω may be connected in series to an amplifier with an output impedance of 40 Ω, and this series circuit can

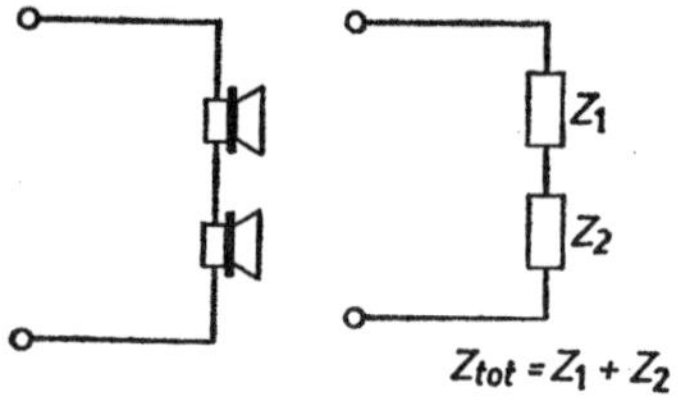


Fig. 48. Loudspeakers in series.

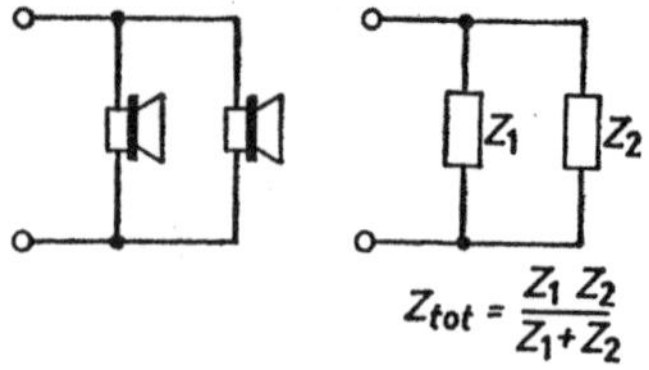


Fig. 49. Speakers in parallel.

provide 4 × 6 = 24 W where the load capacity of the speakers is 6 W. Two speakers of 10 Ω in parallel have an overall impedance of 5 Ω, which matches the amplifier impedance very well. Figs. 48 and 49 respectively show two speakers in series and in parallel, represented by their theoretical circuit diagrams.

**Putting loudspeakers in phase**

When speakers are placed in series or in parallel, they must be in phase, that is to say, their cones must vibrate to and fro in unison. This can be achieved in various ways, and to this end Philips types carry a red mark at the coil input. If, when speakers are connected in series, the red terminal of one is connected to the plain one of the next (see fig. 50), they are in phase, since the current will flow through each speaker coil in the same direction. For parallel connection, the red terminals are connected together (fig. 51). Another way is to follow the cone deflections with the aid of a 4½ or 6 V battery and thus ascertain whether or not the speakers are in phase (fig. 52). They are in phase when the cones both move in the same directions at the same time and out of phase when they do not.

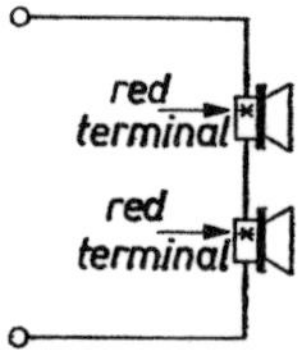


Fig. 50. Correct connection of speakers in series.

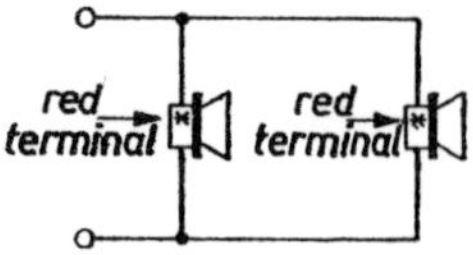


Fig. 51. Correct connection of speakers in parallel.

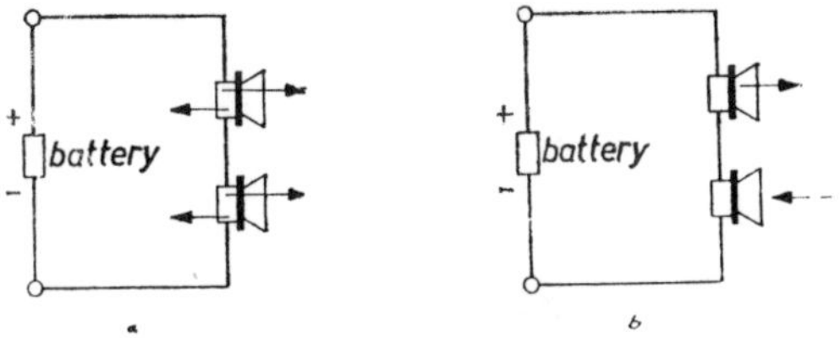


Fig. 52. Putting in phase with the help of a battery; a) the two cones move in the same direction, the phase is correct; b) the phase is not correct.

## Filters

Filters are generally used in monaural reproduction to obtain a "stereo" effect (chapter 5) and for loudspeakers suitable only for a given frequency range, be it high or low. The frequency spectrum is split by means of a filter into two parts in such a way that all frequencies below a certain limit are responded to by one of the speakers and those above it by one or more others. A filter basically consists of two sections connected in parallel, each comprising a coil and a capacitor in series (see fig. 53). The bass speaker is connected in parallel to a capacitor, and the tweeter in parallel to a coil, which allows bass frequencies to pass freely but opposes the high ones. The impedance (a.c. resistance) increases proportionally with the frequency, and hence the impedance is twice as high at double the frequency, and so forth. The same is true in the opposite case.

It is interesting to note that the impedance of a coil does not consist only of a self-inductance, but, on the contrary, also includes a resistance component as a result of the length of wire required to make up the coil. It also

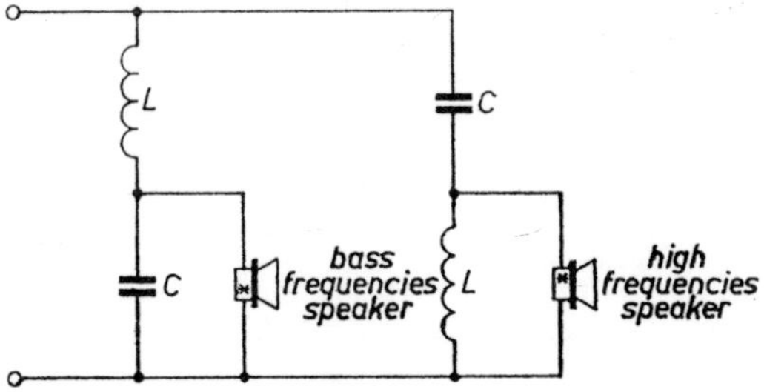


Fig. 53. Connection of speakers through a filter.

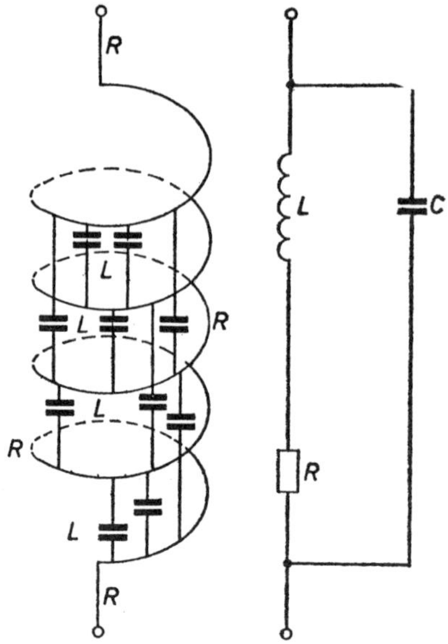


Fig. 54. Symbolic and diagrammatica representation of coil capacitance and resistance.

contains a capacitive component which, however, is virtually negligible for the purpose of the sound installations dealt with here. Fig. 54 gives the actual impedance of the coil both symbolically (on the left) and diagrammatically (on the right). The unit of self-inductance is the henry, abbreviated to H, but the most generally used unit is the millihenry (0.001 H).

The impedance of the coil is not purely inductive, nor can one speak of a pure capacitance in the case of the capacitor. The simplest way of describing a capacitor is as a component consisting of two parallel plates insulated from each other by air or some other insulator. Connecting a battery to a capacitor will produce a charge across it, which will remain across the plates when the battery is removed (fig. 55). If, after the battery is removed, the charge is 1 V,

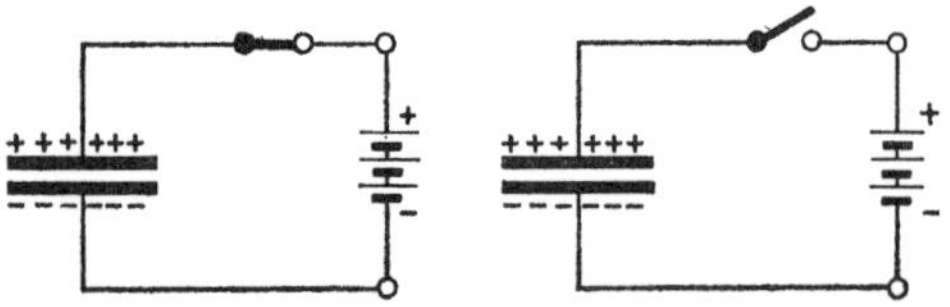

Fig. 55. Capacitor connected to d.c. voltage (left). On the right, the switch is off, but the charge is retained.

and if the capacitor is left to stand, the charge will be found to have dropped to 0.5 V after a while, and will leak away completely in time. One of the decisive factors in the value of the leakage resistance is the out-of-phase character of the polarisation of the dielectric with the charge applied.

Fig. 56 gives the capacitor impedance both symbolically and diagrammatically. The unit of capacitance is the farad, but, because of the great capacitance represented by 1 F, the micro- or picofarad, abbreviated to $\mu$F and pF, and corresponding to $10^{-6}$ and $10^{-12}$F respectively, are the units in general use.

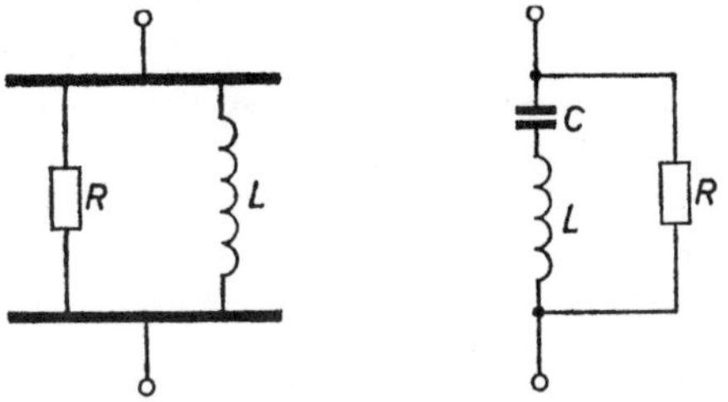


Fig. 56. Symbolic (left) and diagrammatic (right) representation of capacitor impedance.

Fig. 57 is the sound-pressure curve of the filter shown in fig. 53. The attenuation of the signal, i.e. the attenuation of the high frequencies reproduced by the woofer and vice-versa, in the circuits is 12 dB per octave. Figs 58 and 59 show the circuit diagram and sound-pressure curve of a filter with an attenuation of 6 dB per octave.

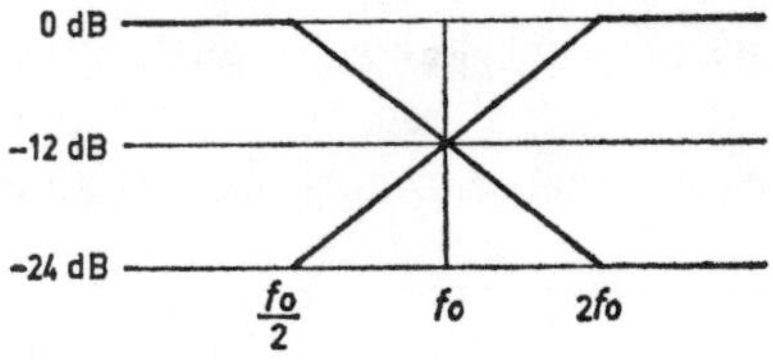


Fig. 57. Sound-pressure curve of 12 dB filter shown in fig. 53.

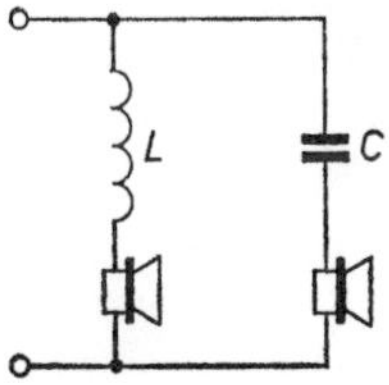


Fig. 58. 6 dB filter.

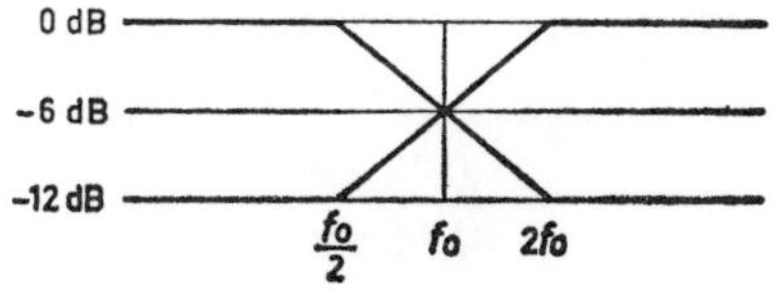


Fig. 59. Sound-pressure curve of filter shown in fig. 58.

The input impedances of the two filter circuits, i.e. those "seen" by the amplifier when the filters and speakers are connected to them, are equal to the speaker coil inductance of the loudspeakers, provided that the latter have the same impedance.

**Filter construction**

If the speaker coil impedance is about 15 Ω or less, the construction of a filter is, in fact, a perfectly feasible "do-it-yourself" job. When high-impedance speakers are used however, with, of course, an equally high-impedance amplifier, the number of turns has to be such that building it at home is precluded. Philips Netherlands of Eindhoven have put this type of coil on the market under code number A3 166 31 expressly for this purpose. The *L* and *C* values of the filter shown in fig. 52 are given in the table below.

| Cut-off frequency | 3.5 Ω | | 5 Ω | | 7 Ω | | 14 Ω | | 800 Ω | |
|---|---|---|---|---|---|---|---|---|---|---|
| | *L* | *C* | *L* | *C* | *L* | *C* | *L* | *C* | *L* | *C* |
| 1000 c/s | 0.78 mH | 32 μF | 1.12 mH | 25 μF | 1.57 mH | 16 μF | 3.15 mH | 8 μF | A316631 | 330 nF |
| 400 c/s | 1.95 mH | 80 μF | 2.78 mH | 62 μF | 3.9 mH | 40 μF | 7.8 mH | 20 μF | | |

Coils for low-frequency filters may be wound on cores of wood, plastics or any other insulating material, but not metal. The best results are obtained with lacquered copper wire 1.2 mm in diameter. The table below gives the self-inductance, number of turns and coil sizes for this gauge of wire.

| Self-inductance | Number of turns | Coil size |
|---|---|---|
| 1.57 mH | 245 | *A* |
| 3.15 mH | 340 | *B* |
| 3.9 mH | 375 | *B* |
| 7.8 mH | 515 | *C* |
| 0.78 mH | 175 | *A* |
| 1.95 mH | 270 | *A* |
| 1.12 mH | 210 | *A* |
| 2.78 mH | 300 | *B* |

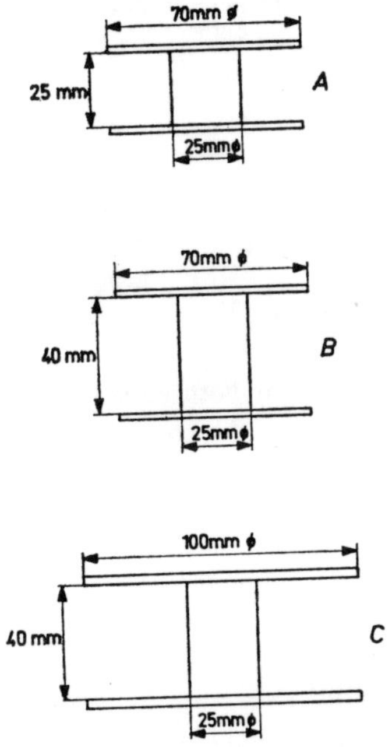


Fig. 60. Core shapes that may be used for winding coils for filters.

Coil sizes are designated *A*, *B* and *C*, an explanation of which will be found in fig. 60.

Paper-insulated (fig. 61), bipolar electrolytic (fig. 62) and ordinary electrolytic capacitors may be used for the filter, but in the latter case, two capacitors connected "alternately" (fig. 63) are used for each capacitance. Two capacitors with these values are often supplied as one unit. In this case, the

Fig. 61. Paper capacitor.

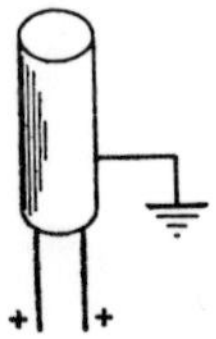


Fig. 62. Bipolar electrolytic capacitor.

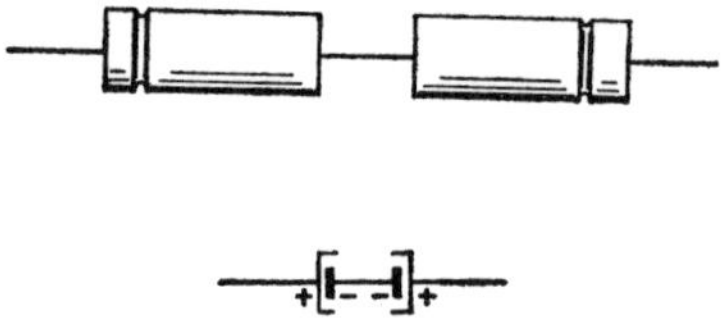


Fig. 63. "Alternate connection" of two electrolytic capacitors.

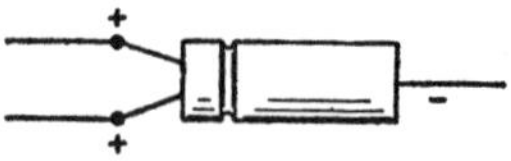


Fig. 64. Two electrolytic capacitors in one housing.

positive terminals are used for both connections and the third tag on the capacitor casing, which forms the connection between them (the negative terminal) may not be used (fig. 64).

CHAPTER 4

# DISTORTION

If often happens that the sound reproduced is not exactly like the original. The reason for this may be distortion or noise, and the former will be discussed first, the latter being left until later on in the chapter.

A difference must be made first of all between two kinds of distortion, namely, linear and non-linear. The latter is caused by the reproduction of sounds not actually contained in the original, whereas linear distortion is the false reproduction of original sound, where the relative intensities of the various notes has been altered. Using as an example the sounds recorded in the studio of a gramophone company on a disc via a magnetic tape, clearly any of the components in the sound installation, including the company's equipment, could be responsible for this distortion. The same applies here as to the "weakest link" of the previous chapter, in that the distortion introduced by a poor-quality amplifier can never be eradicated by the loudspeaker, however good it is.

Next to be discussed is distortion introduced by the speaker and amplifier, which may take three distinct but related forms, namely harmonic, intermodulation and amplitude distortion.

## Harmonic distortion

This is generally produced by non-linear elements or relationships between the various values, and can best be illustrated by means of an example. Fig. 65 shows the "ideal" and actual $I_a$ and $U_g$ curves of an amplifier valve side-by-side, an a.c. voltage being applied across the grid in both cases. The valve with an "ideal" curve will reproduce this voltage exactly, provided that only the amplitude (voltage) is altered. The latter is increased by the amplification of the valve.

In practice, it can happen that the output voltage is equal neither in value nor shape to that applied to the grid of the valve. The distorted output

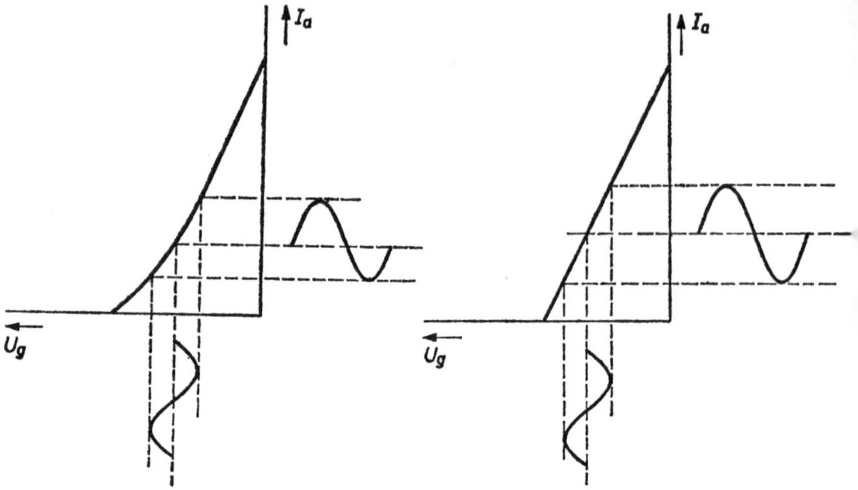


Fig. 65. Amplifier valves with idealised (left) and actual $I_a$-$U_g$ curves. A sinusoidal a.c. voltage is fed to the grids of both valves.

voltage is made up of several a.c. voltages of different frequency and amplitude (fig. 66), and the anode a.c., whose amplitude and frequency are determined by the grid voltage, may be represented thus:

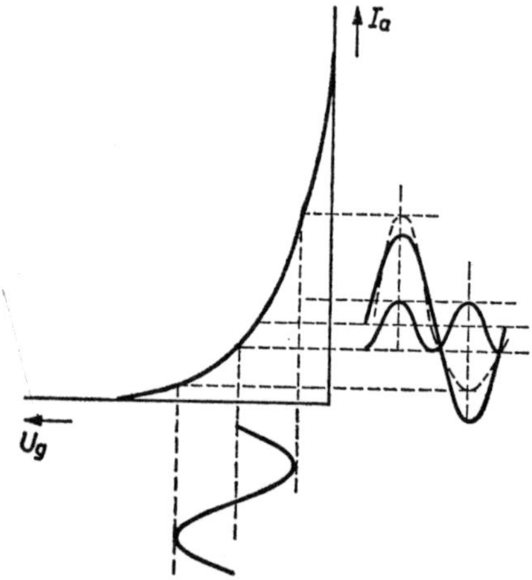


Fig. 66. The distorted anode current, divided into two types of different frequency and amplitude.

$$I_{a\ \mathrm{a.c.}} = A_1 \sin \omega t + A_2 \sin 2\,\omega t + A_3 \sin 3\,\omega t + \ldots + A_n \sin n\omega t.$$

If a sinusoidal voltage of frequency $\frac{\omega}{2}$ is fed to the valve with an "ideal" and hence linear $I_a$-$U_g$ characteristic, the coefficients $A_2$, $A_3$, ..., $A_n$ will be zero, and therefore $I_{a\ \mathrm{a.c.}} = A_1 \sin \omega t$. If, however, this sinusoidal voltage is fed to a non-linear type amplifier valve, the coefficients $A_2$-$A_n$ inclusive will not be zero, thereby producing anode currents whose frequency is a multiple of that of the voltage supplied to the grid. These frequencies are known as harmonics of the original frequency, and their very existence leads to distortion.

To give some idea of the extent of the distortion produced, a "distortion factor" is introduced under the symbol $d$ or $D$ and defined as

$$d = \frac{\sqrt{A_2^2 + A_3^2 + \ldots + A_n^2}}{A_1} \times 100\%$$

Hence, the value of $d$ is determined by the amplitudes of the currents forming the anode a.c. Technical publications on amplifiers, radio sets, etc., give the distortion at a given power, e.g. output power = 10 W, $D < 1\%$.

It is, in fact, impossible for the characteristic of a valve ever to be linear, meaning that there is always the risk of distortion, although there are different methods of keeping this distortion within reasonable bounds; some of these methods will now be briefly discussed.

a. *Feedback*

The circuit diagram for this is given in fig. 67. Part $\beta U_2$ of the output voltage $U_2$ of the amplifier is fed back via a feedback network to the amplifier input.

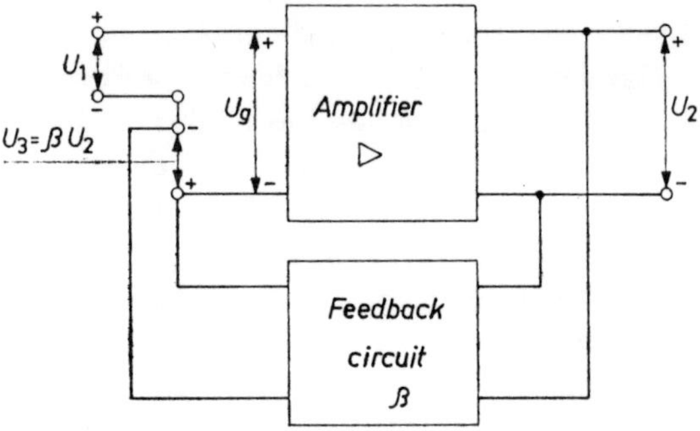


Fig. 67. Theoretical circuit diagram of feedback.

Now the grid voltage $U_g$ is made up of voltages $U_1$ and $\beta U_2$ and is equal to $U_1 - \beta U_2$. Feedback reduces distortion because the higher harmonics are reproduced in very strongly attenuated form, making $A_2, A_3, \ldots, A_n$ very small. There is no need to go into the mathematical derivation here, as it is very complex.

b. *Push-pull or balance circuit*

The diagram of this circuit, containing two valves (triodes) in phase opposition, is shown in fig. 68. During one half-cycle of the grid a.c. voltage, the grid of valve I, for example, becomes less negatively charged, causing an anode current to flow through that valve, while the same applies to valve II in the second half-cycle. Both currents will, however, flow through the primary winding of the transformer and, by virtue of their alternating nature, induce a current in the secondary chain. All this is shown symbolically in fig. 69.

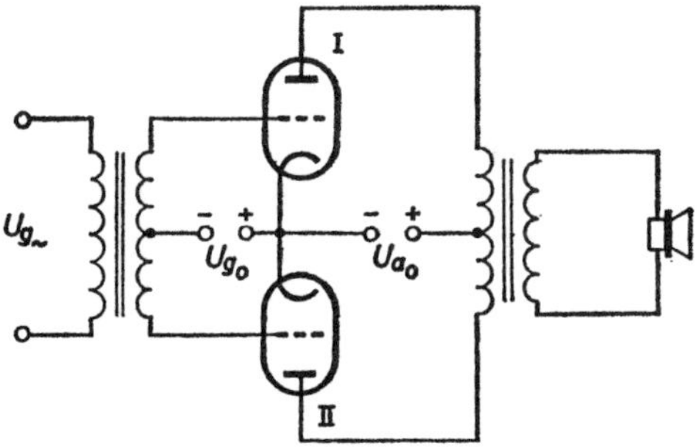


Fig. 68. Theoretical circuit diagram of the push-pull or balance circuit

The advantage of a balance circuit is that all even harmonics are suppressed. If the valves are identical, however, the positive and negative parts will differ in sign and shape, and only the odd harmonics will be included.

Having examined the amplifier to see how harmonic distortion is introduced, let us find out how the speaker can give rise to it.

It can do so in two ways. A look at the graphic representation showing the deflection of the speaker coil from the state of balance against the force acting upon the coil shows that, in general, the relationship between these two factors is linear only when the deflections are slight. Such a non-linear relationship results in the introduction of odd harmonics when a sinusoidal input voltage is applied. The other cause of distortion can be the inhomo-

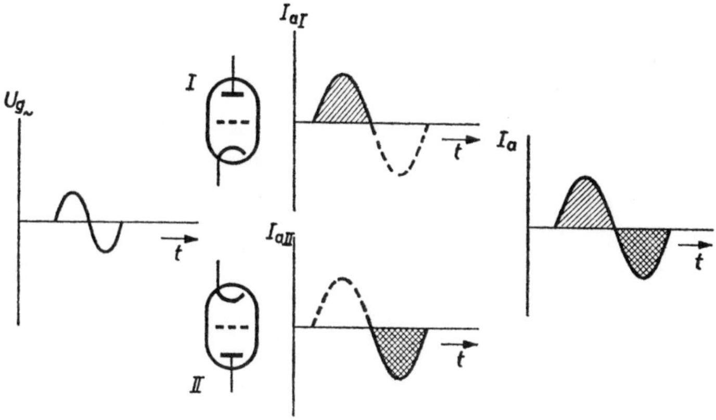


Fig. 69. Diagrammatic representation of the operation of the balance circuit

geneity of the magnetic field, and this type of distortion occurs especially when the speaker coil is taken partially out of the field, and, as already mentioned in chapter 1, this can be prevented by making the height of the coil smaller than the size of the air-gap.

**Intermodulation and amplitude distortion**

Intermodulation distortion can occur when two frequencies are fed to a speaker simultaneously. If these frequencies are very close together, the distortion is called amplitude distortion. The former type is of importance only when a very low frequency, in the region of the resonant frequency, is fed to the coil in combination with a higher frequency. Intermodulation distortion can be very greatly reduced by strongly attenuating the resonant frequency or by the use of several loudspeakers divided up into tweeters and woofers with the aid of a crossover filter. Amplitude distortion becomes most noticeable when the difference in frequency is 1000 c/s.

This chapter has only really skated over the surface of the subject of distortion, but this has been done deliberately, as all that was intended was to give an idea of some of the difficulties with which the loudspeaker designer has to deal. Together with the information given in chapter 1, it will show that designing a loudspeaker is not exactly child's play, since there are too many requirements, desires and possibilities to be taken into account.

## Noise

The amplification of a radio receiver or amplifier cannot be increased indefinitely without some adverse results. If the volume control of an ordinary radio set is turned full on, a crackling and hissing noise is heard, also known technically as noise. It may be caused by many things, both inside and outside the set. External factors include noise from motor-cycles, cars, vacuum cleaners, etc., and also atmospheric disturbances; it can even be of cosmic origin. Within the set, resistors, capacitors, valves, etc. may also produce noise, divided up into thermal and excess noise and hail effect, but this subject will be left at this point.

# CHAPTER 5

## STEREOPHONIC AND MONAURAL LOUDSPEAKER INSTALLATIONS

There are two types of sound-reproduction installations, stereo and mono. In the latter, all the speakers render the same sound (fig. 70), whereas in a stereo layout, on the other hand, the sound provided by the speakers can be different for each one (fig. 71).

Both these systems, stereo (*b*) and mono (*a*), are illustrated in fig. 72. A mono-generator unit, say, a pickup head, provides only one voltage, which is fed to the input of the mono-amplifier and amplified, and the speakers connected to this amplifier all receive the same signal. A generator unit designed for stereo use, however, supplies two generally different voltages, which are delivered to the stereo amplifier composed, in effect, of two mono-amplifiers,

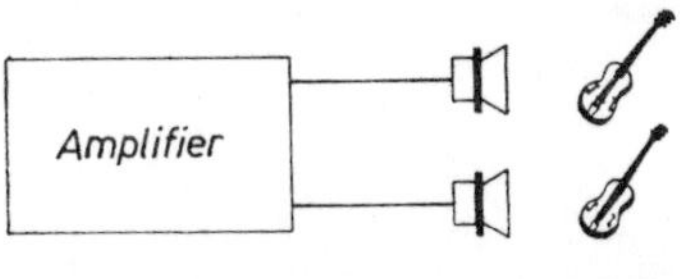


Fig. 70. Symbolic representation of a mono-amplifier.

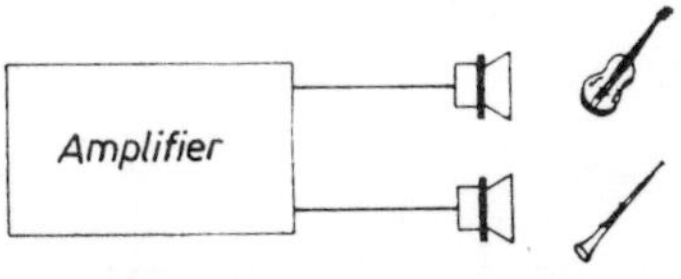


Fig. 71. Symbolic representation of a stereo-amplifier.

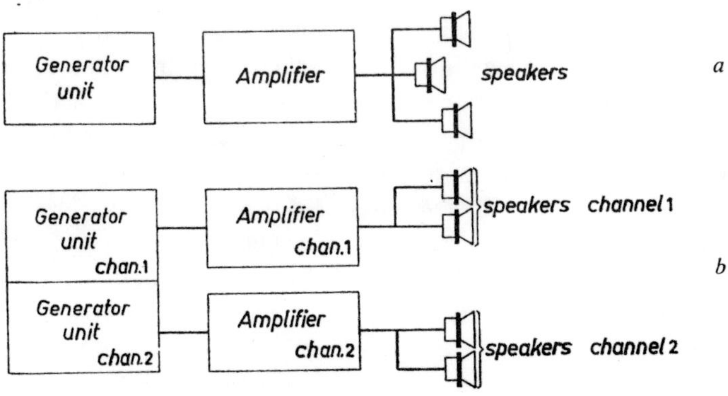


Fig. 72. Diagrammatic representation of a mono- (*a*) and a **stereo-installation (*b*)**.

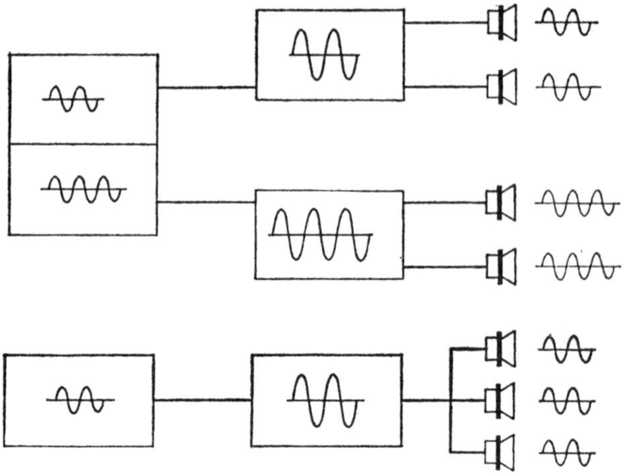

Fig. 73. Voltages in a mono- (*a*) and a stereo-installation (*b*).

whence they are fed to the speakers. As the voltages supplied to the amplifiers may differ, so also may those delivered to the speakers. For a diagram of the possible voltages in stereo and mono-installations, see fig. 73.

Let us now deal with the positioning of the speakers.

**Loudspeaker layouts for stereophonic installations**

For stereo reproduction, the speakers should be placed in such a way that the particular properties of this system are used to the full. The prime requirement made of stereophonic reproduction is that it shall give an accurate sound-picture of the location of the instruments in an orchestra, without any perceptible "gaps". Thus, besides having a "localised" character, i.e. putting each instrument in its proper position, the sound-picture must be continuous. To aid in this localisation, it is also necessary for the acoustic properties of the concert hall, such as reflections and echo times (see chapter 6), to be reproduced on the gramophone record. Unlike mono-installations, where tweeters are used for treble response, this sort of reproduction is out of the question for stereo layouts, as the stereo effect would be blurred.

Stereo reproduction is a system whereby the sound is rendered as though it emanated from a wide source. It would be manifestly ridiculous to introduce a complete concert hall stage into an ordinary sitting-room measuring twelve feet by twenty-one, and therefore the sound is recorded from a base of no more than about eleven feet, making it possible to keep the proportions as actually used during recording down to private house size. The speakers in a stereo-installation, therefore, should be placed about eleven feet apart to obtain a faithful reproduction of the original sound-picture.

Obviously, the positioning of stereo loudspeakers will vary according to the dimensions and shape of the room. Since the sound-picture depends to a great extent upon the distance between the speakers, it would be as well to discuss some of the different ways in which they may be arranged in rooms of various sizes.

*Large rooms*

The distance between the speakers and the side-walls is very great, and therefore direct reflection will be reduced to a minimum, so that the sound-picture will be determined almost solely by the direction of the axes of the speakers themselves. The point of intersection of these axes, known as the radiation axes, is about five feet in front of the listeners. The position of this point *C* may be shifted along line *DE* by varying the angles between the radiation axes and the side-walls *X* and *B*, thus allowing the sound-picture to be suited to the acoustic properties of the room. These arrangements are shown in fig. 74, in which the cross-hatched area *FGKH* gives the part of the room where the best stereo impression is obtained, which may include 70% of the space available.

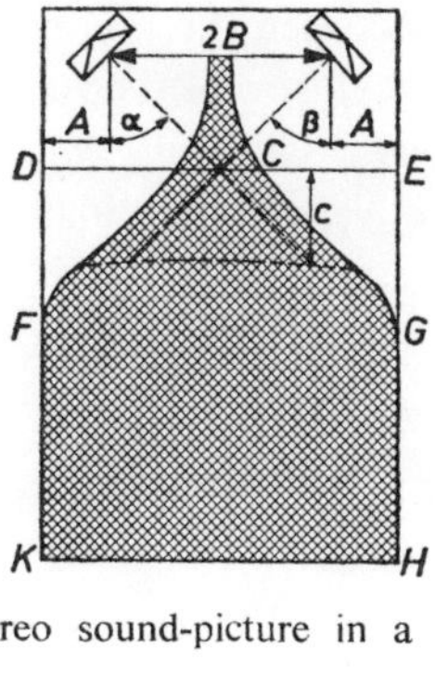


Fig. 74. Stereo sound-picture in a large room.

*Medium-sized rooms*

In this type of room, *A* will be the same size as or smaller than *B*. Here, reflection from the walls may give a great deal of trouble, but it can now be reduced by applying sound-deadening material in the immediate vicinity of the speakers, or by altering the radiation patterns of the speakers so that there is hardly any sound-radiation in the wrong direction. This is very difficult to do, however. Fig. 75 shows the arrangement of the speakers, the cross-hatched area *FGHK* illustrating the zone where the best stereo sound-picture is obtained.

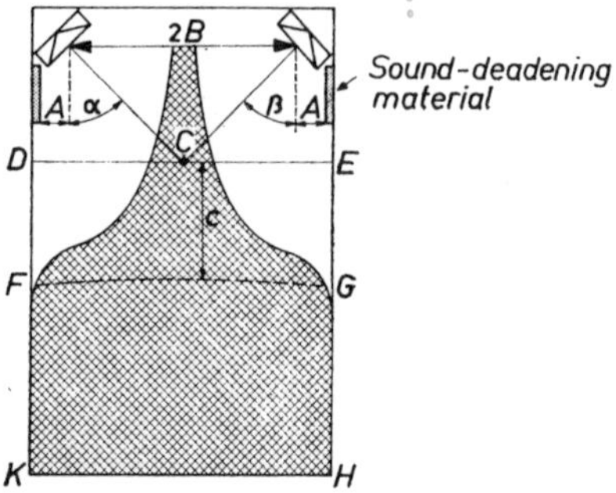


Fig. 75. Stereo sound-picture in a medium-sized room.

*Small rooms*

These include rooms where *A* is much smaller than *B*, whilst 2*B* = 3–4 yds approximately, and therefore *B* is about five or six feet. If the speakers are placed in the corners of the room, the sound-base will not be more than about five feet, which is too short for proper stereophonic reproduction. Because the time-difference between the original and the reflected signal is imperceptible owing to the small size of the room, direct reflections may be used to widen the sound-base. The wall, which now serves as a sound-reflector, must therefore be covered with a material that reflects as much of

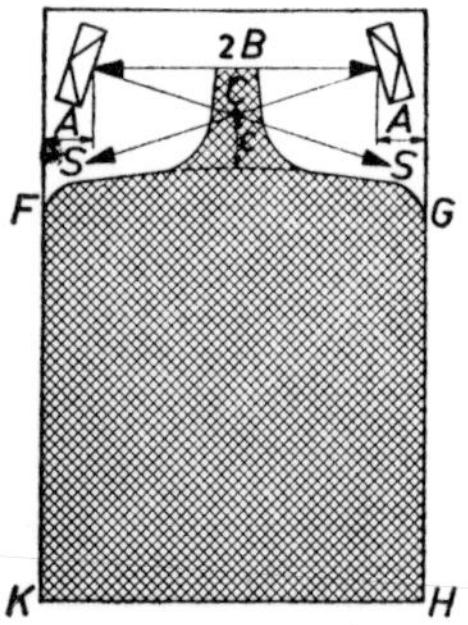


Fig. 76. Stereo sound-picture in a small room.

the sound as possible, e.g. plaster. The broader radiation thus produced and the imaginary sound-base give an effective stereo-picture covering about 50% of the total floor area (the cross-hatched zone in fig. 76).

*More loudspeakers*

A better sound-picture can be obtained by the use of more loudspeakers, e.g. two woofers, two middle-register speakers and two tweeters, although the problems raised by this arrangement are too complex to be explained in general terms.

## Loudspeaker arrangements for mono reproduction

When setting up speakers for mono reproduction, it is important to know how to obtain a "stereo effect". The term "diffuse sound" has already been mentioned in the earlier part of the chapter, and is, indeed, the most important factor for general purposes. It may be obtained by several different methods, the simplest being to arrange the speaker in such a way that the sound radiated from it impinges at random on reflective surfaces. This method is most effective with concave surfaces (fig. 77), although diffusion will also be produced by sound striking a flat surface at a fairly wide angle of incidence (fig. 78). A more complex form of diffusion is obtainable by making

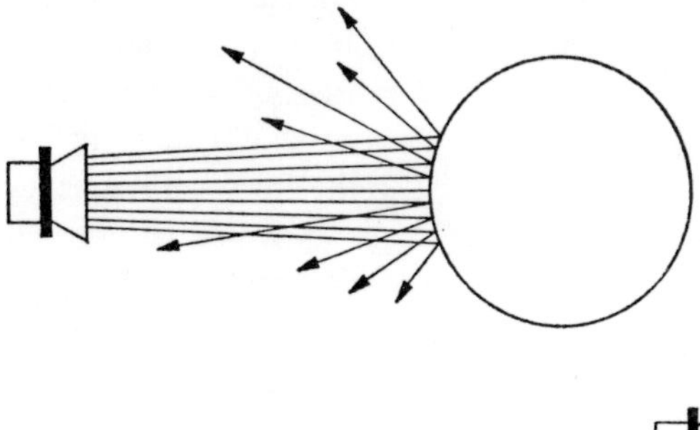

Fig. 77. Diffusion of the sound by reflection against curved surfaces.

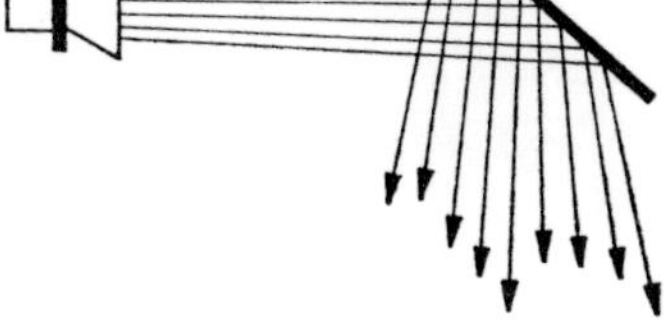

Fig. 78. Diffusion by reflection from flat surfaces.

only the treble register diffuse. If a crossover filter is used, all frequencies higher than the crossover frequency are responded to by the tweeters, and they can be diffused by placing the tweeters at various points around the room. The reason for diffusing these high frequencies in particular is their directional properties. Now, frequencies below 500 c/s do not play a very important part in our sense of directional hearing. In fact a man put blindfold into a room where all the sound was at 500 c/s or below would have great difficulty in telling the direction of the sound, whereas it would be extremely easy if its frequency were about 2000 c/s, for example. The main reason for this lies in the speaker itself, since it radiates the higher frequencies in beam form: the higher the frequency, the stronger the beam (in the radiation pattern). This is not so for the lower frequencies, however. Fig. 79 shows three speakers divided by a crossover filter tuned to 500 c/s into two groups, one speaker for the range below 500 c/s and two for the frequencies above this.

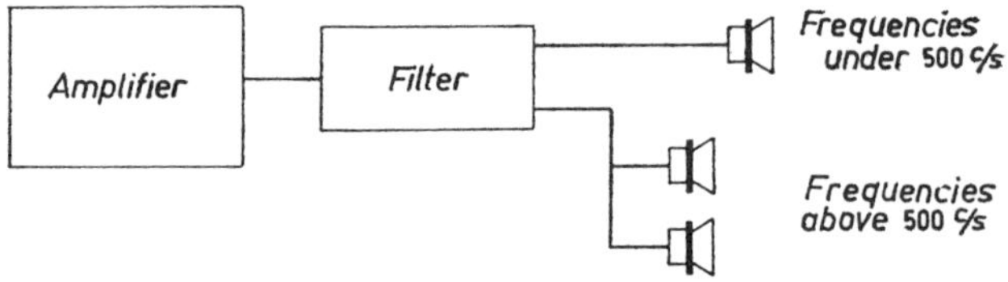


Fig. 79. Amplifier with filter.

In theory, any crossover filter could be used, but, since our sense of directional hearing begins to come into effect at 500 c/s, peculiar things would happen if the crossover frequency were taken at, say, 1200 c/s. One of these is "jumping". It can happen in a piece of music that a frequency of 700 c/s is immediately followed by one of 1600 c/s. If the crossover frequency were 1200 c/s, this would mean that the lower frequency would be radiated from the woofer and the other by the two tweeters, and since both frequencies are "directional", the sound would appear to jump from one speaker to the others, thus spoiling the listener's enjoyment. A crossover at 500 c/s would largely avoid this.

CHAPTER 6

# INDOOR ACOUSTICS AND SOUND-REPRODUCTION IN THE OPEN

## Indoor acoustics

Let us now turn our attention to the acoustics of an ordinary room, where a good speaker is also very important to the quality of the sound. One of the important factors to note here is that even minor alterations in the arrangement of furniture and soft furnishings can greatly affect the acoustics of the room, and judicious use of them can, therefore, lead to better acoustic effects. This will be discussed in greater detail on the next few pages, but it would be as well first of all to explain a few acoustical terms and phenomena.

## Indirect sound

Sounds heard in the open seem different from the same sounds when heard within the confines of a room. In the former case, the sound energy is radiated outwards in all directions and only a small part of it will be heard directly (fig. 80). Indoors, on the other hand, not only the direct sound, but also a high proportion of the reflected sound sent back from the walls is

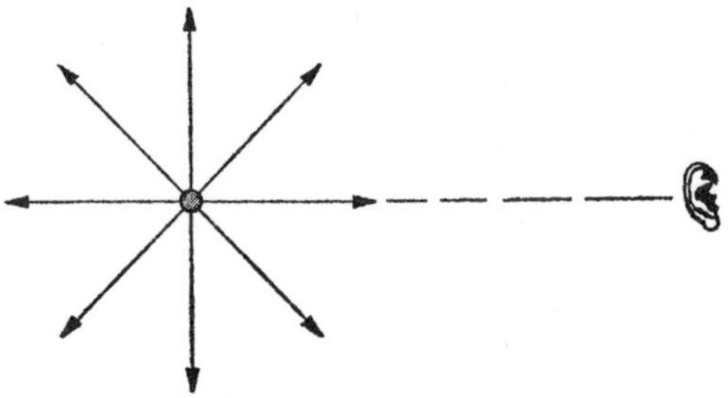

Fig. 80. Point sound-source. Only a part of the energy will reach the ear.

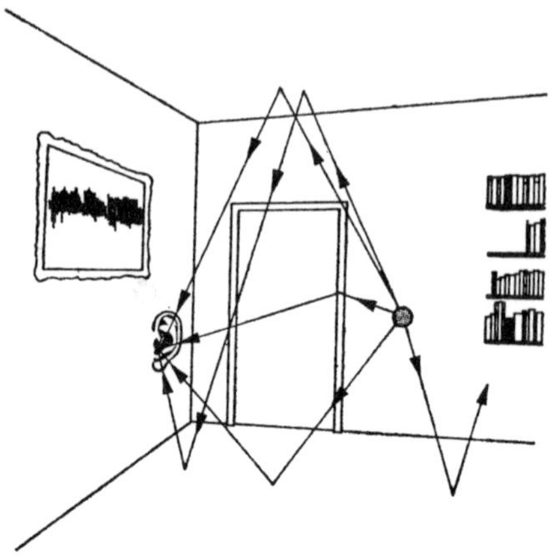

Fig. 81. Reflection of the sound-waves from the walls makes an important contribution towards the sound-picture inside a room.

heard (fig. 81). The latter type is known as indirect sound, and, as it has further to travel than direct energy, it will reach the ear later and thereby increase the total intensity of the sound (fig. 82). The total sound-intensity at any one point, therefore, is the sum of the direct and indirect radiation, which is why the density or concentration of energy is greater in a closed room than in the open under the same circumstances. Some of the energy will be absorbed by the wall, and even more by objects present in the room, the extent to which this happens being governed by the absorptive properties

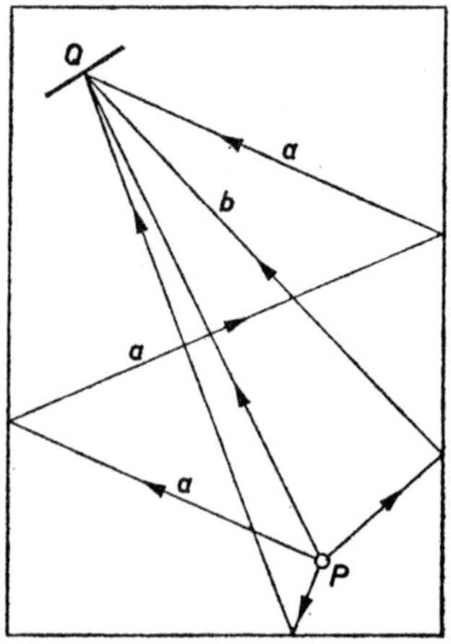


Fig. 82. Reflected sounds *a* and *b* have to travel further to get from *P* to *Q* than the direct sound.

of the material of which the objects are made. However, in spite of the fact that the sound reaches us from a variety of directions and with different intensities, we can still distinguish its original direction because it is the direct sound that reaches the ear first.

### Echo and echo time

When a sound-source is suddenly cut off, only the indirect component of the sound is heard, its intensity dropping off sharply according to the dimensions and absorptive properties of the room, and this sound is called the echo. The internationally accepted standard echo time is the time taken for the intensity of the sound to fall off to one millionth of its original value, corresponding to a reduction of 60 dB in the level (fig. 83). This echo time is one of the factors determining the acoustic quality of a room. Too long an echo time tends to render the sounds blurred or indistinct, since they overlap. 0.5 seconds is a good echo time for an ordinary room, whereas it will often be as much as two seconds in a concert hall.

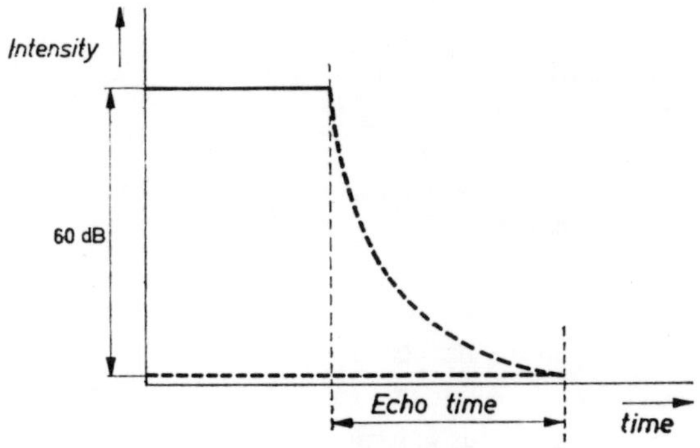


Fig. 83. Graphical representation of the fall-off in sound-intensity and the definition of echo time.

The importance of indirect sound has been sufficiently emphasised by the foregoing. As it is produced by the room in which the sound is made, it is obvious that the room's acoustics determine its suitability for the purpose, and it must, therefore, be regarded as an essential part of the sound-source. The echo time of the room has also been discussed, although it has not, in point of fact, been really accurately defined, since it varies according to the frequency or frequency range of the sounds. Moreover, the echo is also affected by the shape of the room, its furniture and the reflective or absorptive properties of the soft furnishings and objects in it, as well as the actual layout of these objects.

### The shape of the room and the resonant frequency

The resonant frequency has already been discussed in connection with the loudspeaker, and it also has to be taken into account in the acoustics of a room. The air in the room, in fact, has a number of resonant frequencies determined by its shape, and these are easily calculated for a rectangular room. For example, in a room about sixteen feet long, twelve feet wide and nine feet high, the number of resonant frequencies will be found to be very high, and those up to about 170 c/s are shown in fig. 84 by way of illustration. Resonance does occur at frequencies higher than 170 c/s, but such resonant frequencies are so numerous and close together that it is impossible to give a comprehensive diagram of them.

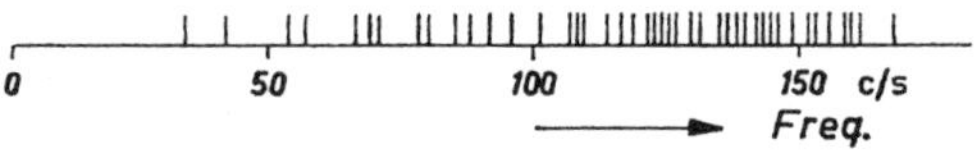


Fig. 84. Resonant frequencies in a room measuring 5 × 4 × 3 metres: only the frequencies below 170 c/s are measured.

When the echo begins to be heard after a sound-source that produces several frequencies has been cut off, the resonant frequencies of the room will predominate strongly. This is logical, but nevertheless annoying, and is made all the more complex by the fact that the attenuation is not generally the same for all frequencies. These difficulties are most pronounced at lower frequencies, since the resonant frequency or frequencies vary widely.

An acoustically ideal room has an infinite number of resonant frequencies, all of which are attenuated in the same way. This, however, is hardly a practical proposition, although it is possible to prevent the frequencies from lying too close together and to obtain a more even distribution of them in that part of the spectrum where it is of real importance, i.e. in the lower frequencies. For this purpose, the room is preferably irregular in shape; rectangular rooms in general, and cubical ones in particular, should be avoided. The proportions of the room are not so critical, provided that none of the dimensions is equal to or a multiple of any of the others. Extreme dimensions, such as obtain in very long or narrow rooms, should be avoided. The walls of recording studios are often "saw-toothed" to provide the necessary irregularity, and this can be copied at home by the judicious use and location of pieces of furniture, pictures, etc., although, to be effective, an object must have dimensions of the same order of magnitude as the sound wave-lengths. The relationship between the wave-length and frequency of a sound vibration is given by the formula $f = 330/\lambda$, where $f$ and $\lambda$ represent the frequency (in c/s) and wave-length (in metres) respectively. The most general formula for this relationship is $f = v/\lambda$, with $v$ representing the rate of propagation of the sound. The first formula is obtained because here we are interested only in the rate of propagation of sound in air, which is 330 m/s. A sound with a wave-length of one metre, therefore, will have a frequency of 330 c/s, wave-lengths of two and three metres giving frequencies of 165 and 110 c/s. Since the wave-length increases as the frequency drops, only large objects will affect low-frequency sound.

### Diffuse sound

A further advantage of giving the room an irregular shape and placing objects in it is that reproduction becomes diffuse, giving an impression of space (see also page 49, "Loudspeaker arrangements for mono-reproduction"). Large, concave surfaces, however, combat diffusion, since they concentrate the sound (fig. 85), whereas convex surfaces aid diffusion (fig. 86).

The foregoing shows that the acoustics of a room in a dwelling-house are affected to no slight extent by the furniture and soft furnishings. Of course, special absorbent materials and constructions can be used, and those

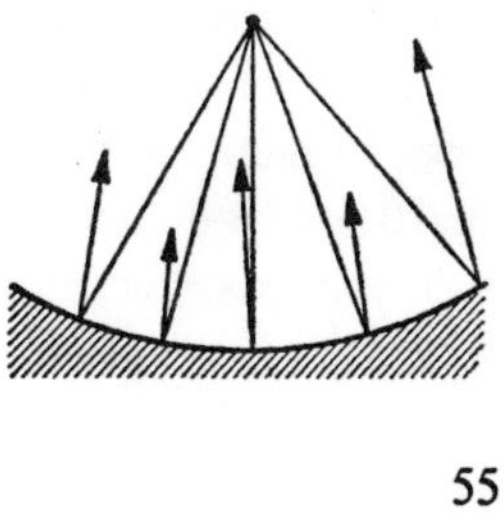

**Fig. 85. Diffusion is combatted by concave walls and ceilings.**

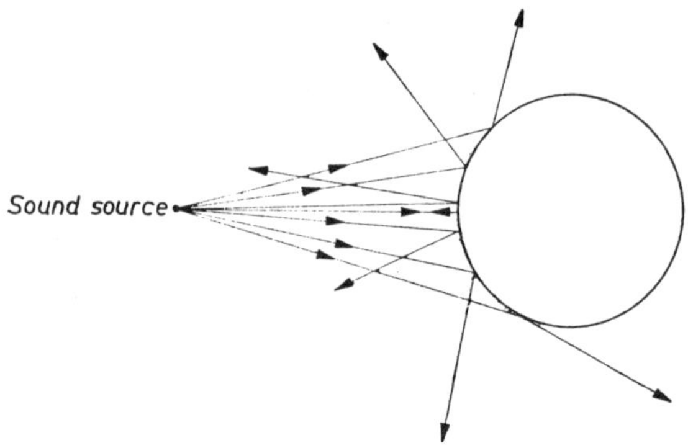


Fig. 86. Diffusion is increased by convex surfaces.

specialising in this field will doubtless make full use of this. It hardly seems necessary, for practical purposes, to go into this subject here in greater detail. The effect of the abovementioned materials and constructions is based on special absorptive properties that have already been dealt with.

Let us then conclude with a few hints arising from the theory discussed in this chapter.

1. Break up large wall areas with chests of drawers, desks, etc.
2. Disrupt the normal parallel lines formed by ceiling and floor with the furniture in the room.
3. Avoid having too many heavy carpets and curtains.
4. Avoid having one highly absorptive area, e.g. large, heavy curtains, when the rest of the room is acoustically hard.

**Sound-reproduction in the open**

That there is an apparent difference between sounds heard in the open and the same sounds heard indoors has already been mentioned earlier on in this chapter. It would be logical to conclude from this that the power of an amplifier intended for the reproduction of sound in the open has to be higher than that of one designed purely for indoor use, but there is yet another factor to be considered, namely the ambient noise-level, which may be 10–20 dB higher outdoors than inside a room.

There are other interesting aspects to the problem, moreover, one of these being "keyhole effect". Since there are no walls to reflect the sound, it is not diffused at all, and the sound reaches the ear in highly concentrated form All of it seems to come directly from the loudspeakers, as indeed it does, giving the impression of having one's ear pressed against a keyhole and of receiving all the sound through it. This effect can be reduced by using several speakers, thus also reducing the concomitant psychological irritation resulting from the fact that the abovementioned effect causes any distortion in the sound to be magnified. This is because there are no diffuse sounds to distract the attention, thus allowing the listener to concentrate entirely on the direct component. A simple test of this is to play a gramophone record that crackles. Indoors, the crackling passes almost unnoticed, but becomes really irritating when heard out-of-doors.

An effect that can be both useful and annoying is echo, which may be compared to indirect sound, and can be produced by sound-reflection from the edge of a wood, sides of buildings, etc. If the echo time (fig. 87) is less than 0.1 second, it can make the sound "come alive", but times of more than 0.1 second are usually annoying and tend to make the spoken word unintelligible.

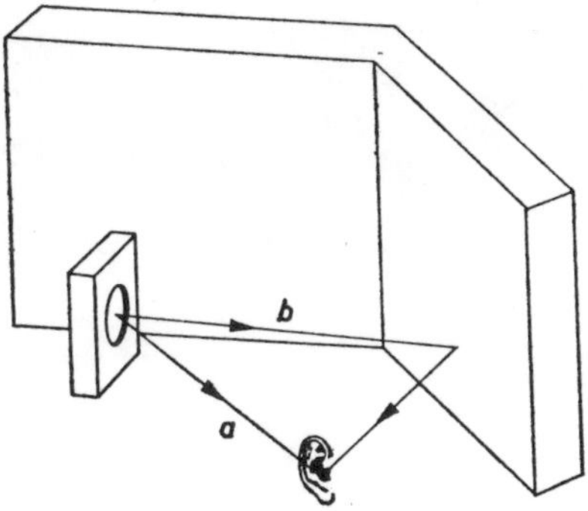


Fig. 87. The time-difference between direct and reflected sound is determined by the difference in the distance that both sound-waves have to travel. *a* is the direct and *b* the reflected sound.

If desired, a loudspeaker box may be placed in the open without modification, provided that the leads are long enough, but to weatherproof it, and especially to protect it against damp, a large plastic bag or shaped plastic

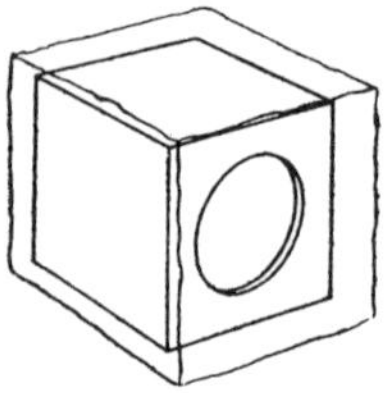

Fig. 88. Loudspeaker box protected against the weather by a thin plastic sheet.

sheet may be wrapped around it. This material provided sufficiently thin, has no acoustic effect, and efficiently protects the speaker and its box (fig. 88).

Fig. 89 illustrates the recommended arrangement of the speaker in such a situation.

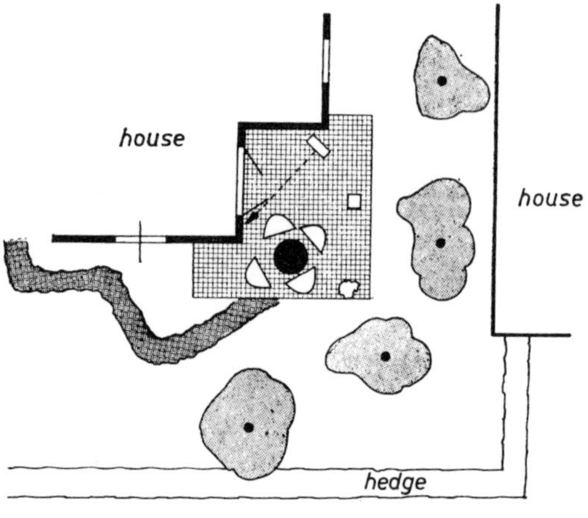


Fig. 89. A typical speaker layout.

CHAPTER 7

## THE PRINCIPLES OF BAFFLES AND LOUDSPEAKER BOXES

Chapter 1 discussed the principles on which a speaker works, and how the a.c. fluctuations fed to the speaker coil are transformed by the speaker cone into vibrations that are audible provided that their frequencies lie within the range of 16 to 16,000 c/s. The air-vibrations produced by the movements of the cone will now be examined more deeply.

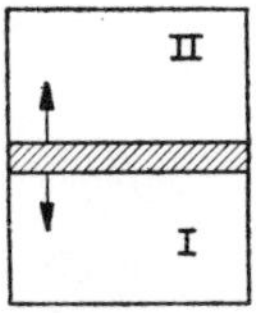


Fig. 90. The "piston and cylinder" analogy.

Fig. 90 shows an airtight piston working in a cylinder. If the piston is pushed downwards, the pressure in chamber I will increase, while that in chamber II will decrease, the reverse happening when the piston is pushed the other way. The speaker cone may be compared to the piston and the air behind and in front of it to that contained in chambers I and II, so that, when the cone moves forwards, the pressure in front of it will increase, or, in other words, the air will become denser, whilst a slight depression will be created behind it. Similarly, the reverse happens when the cone moves backwards. There is, however, an essential difference between the cone and its "piston and cylinder" analogy, in that the air compressed by the piston cannot escape and remains shut up inside the cylinder, whereas the air "compressed" by the speaker cone has infinite room in which to expand. Although, because of this large space, it would be nonsensical to refer to air-compression in connection with the speaker cone, therefore, there is, nevertheless, a layer of denser air just in front of it, and this layer may be regarded as a surplus of air-molecules per unit volume, which will now be passed on from one space to the next, thus propagating this local air-densification (fig. 91). Where the air is attenuated, however, i.e. where a slight depression is created, there will be a deficiency of air-molecules, which will be made up

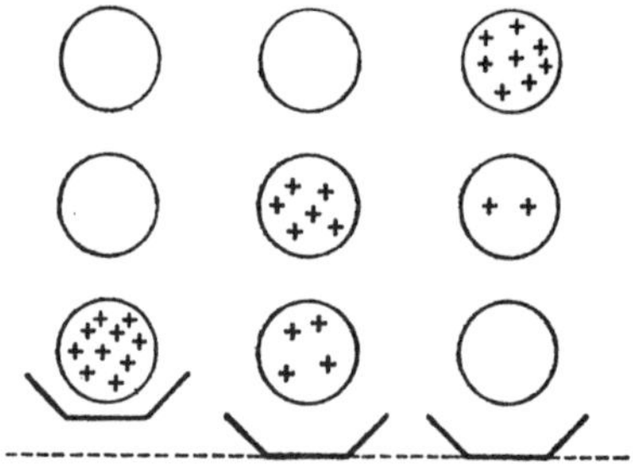

Fig. 91. Diagram of the propagation of sound-waves.

from the neighbouring volumes of air, this process continuing ad infinitum, thus propagating the air attenuation as well.

One may well ask the question "What happens when attenuated and denser zones meet?" The answer depends on their relative "sizes" or values. A highly dense zone, with a large surplus of air-molecules, will not only neutralise a slightly attenuated zone, but will still leave a surplus. Where the two types of zone are equal, they will cancel each other out. This actual process is, in fact, much more complex than this, but the foregoing does demonstrate the basic concept, and it is time that the use of a baffle were discussed.

## The baffle

The simplest kind of baffle is merely a circular board with a hole in it, behind which the speaker is placed, fig. 92 showing a speaker mounted on a baffle of radius *a* metres. The denser air-zones are indicated by a plus and the slight depressions by a minus sign, and it will be seen that an "acoustic short-circuit" can be produced if, when the vibrations emanating from the back of the cone travel around the edge of the baffle to the front, partial or total neutralisation of the denser and attenuated zones occurs. This can very

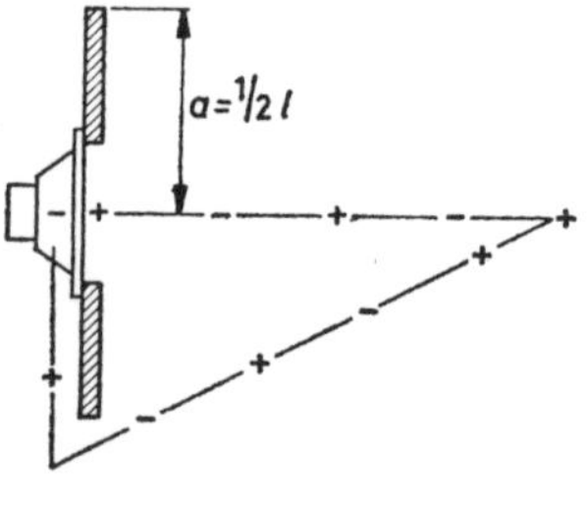


Fig. 92. Baffle with speaker

easily happen if the radius *a* of the baffle is small, meaning that the vibrations from the rear of the cone lag only slightly. Attenuation is much less marked when there is a certain time-difference between the two types of vibration. The cut-off wave-length of a baffle, i.e. that at which attenuation just does not occur, is 21 metres. Hence, the cut-off frequency may be calculated with the aid of the equation: cut-off frequency = 333/21.

Below this frequency, however, the mutual attenuation of the sound-waves will, indeed, become apparent, and is clearly shown by the sound-pressure curve of the baffle, decreasing at the rate of 6 dB per octave below the critical frequency. It is diagrammatically shown in fig. 93.

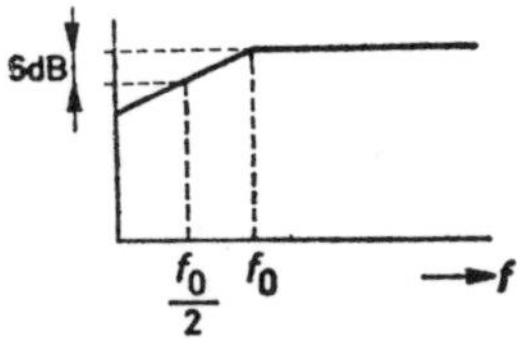


Fig. 93. Sound-pressure curve of a baffle.

It has already been stated in chapter 1 that the sound-pressure curve of the speaker falls off at a rate of 18 dB per octave below the resonant frequency (fig. 94). It would, therefore, be interesting to take a closer look at the appearance of the sound-pressure curves of the speaker/baffle combination at various critical resonant frequencies. In order to give a clearer idea of this, fig. 95 shows nine curves, those marked $a_1$, $b_1$ and $c_1$ being those for speakers mounted on an infinitely large baffle. Figs. $a_2$, $b_2$ and $c_2$ show diagrammatic baffle-curves, the kink in the lines indicating the position of the critical frequency. Finally, the results of the combinations ($a_1$ $a_2$), ($b_1$ $b_2$) and ($c_1$ $c_2$) are shown in figs. *A*, *B* and *C*.

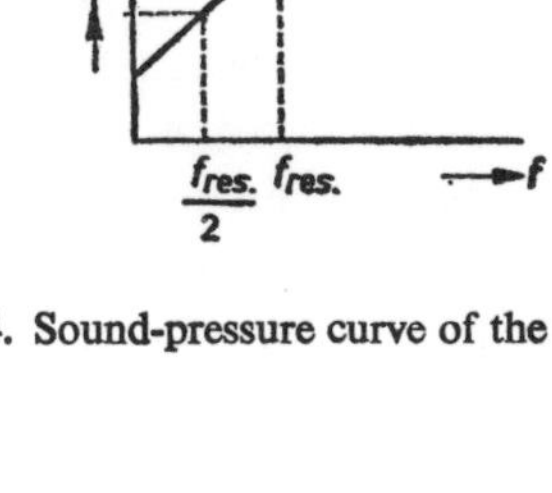


Fig. 94. Sound-pressure curve of the speaker.

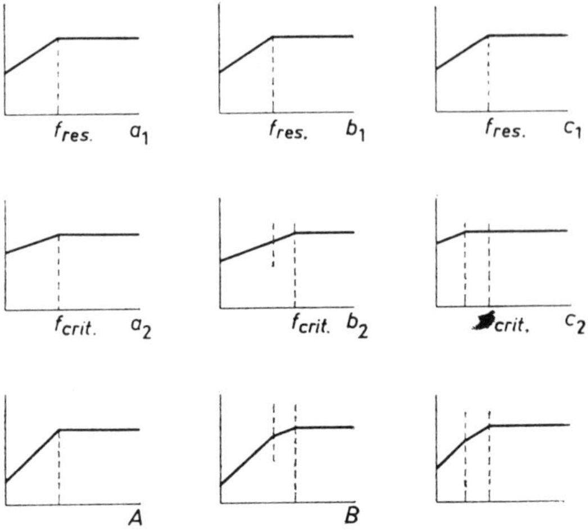


Fig. 95. Sound-pressure curve of speakers mounted on baffles with different critical frequencies.

## Various types of baffle

Baffles may be of various shapes and sizes, viz:

A. The "*normal*" *baffle*, consisting of a board with a hole in it, behind which the speaker is placed. The example chosen here is circular, for the sake of simplicity, although this shape gives a very sharply defined critical frequency. "Baffle dip" is often mentioned in this connection, and it can be very annoying. It is therefore better to use, say, an asymmetrical shape (fig. 96).

Fig. 96. Simple baffle construction.

B. The *infinite baffle*. As the name implies, this type is generally very large, but need not invariably be so. Large dimensions are, indeed, necessary to allow as much bass vibration as possible to be propagated, i.e. to avoid acoustic short-circuiting. At the same time, the same effect can be attained by the use of a smaller baffle with a lower critical frequency than the resonant frequency of the loudspeaker, which will explain the apparently illogical use of the word "infinite" in some cases. Speakers are sometimes built into walls to obtain as large a baffle surface as possible, the vibrations produced by the rear of the cone being absorbed by suitable material (fig. 97).

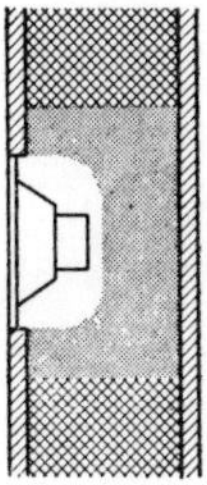

Fig. 97. Speaker built into a wall to obtain an "infinite baffle"

C. Fig. 98 shows a variation on the "normal" baffle, where the distance between the edge of the baffle and the axis of the speaker is increased by a projecting rim. In such a case, length *b* in the figure may not be more than half length *a*, or resonance will occur.

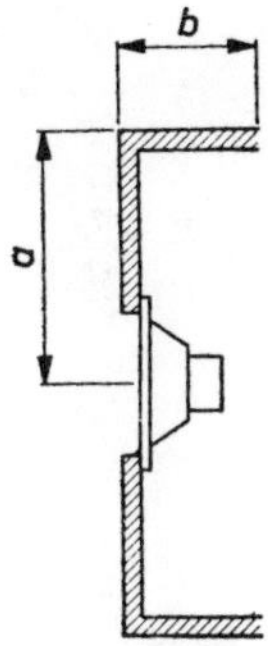


Fig. 98. Baffle with projecting ed ges.

It will be obvious from the foregoing that a baffle is vitally important to good bass response, but large baffles have long critical wave-lengths and hence low cut-off frequencies. The use of very small baffles gives high cut-off frequencies and bass notes will not be heard owing to acoustic short-circuiting, and so, in order to obtain a useful cut-off frequency, the baffle must be of reasonable size. The largest dimension *a* of a baffle, measured from the axis of the speaker, should be 1.2 metres to obtain a cut-off frequency of

70 c/s. The resultant baffle, however, has a diameter of 2.4 metres, and is, therefore, unsuitable for use in the home, necessitating the use of smaller, specially shaped types (see model 1, page 92). Their dimensions are, nevertheless, rather large. To get around this, various types of loudspeaker box have been developed, the best-known being the acoustic box.

## The acoustic box

In theory, the acoustic box consists of a completely closed box with a baffle aperture in one wall (fig. 99). It is by no means small, but it is still smaller than an equivalent bass baffle. Inside, it is lined with pieces of sound-absorbing material held in place with laths. The vibrations produced by the back of the cone die away inside the box, and, if there were no means of absorbing this energy, the loudspeaker would give unsatisfactory performance. In order to attenuate this unwanted sound as much as possible, the absorbent sheets are not applied directly to the walls of the box, but stand off about twelve to thirteen millimetres away from them (fig. 100). Suitable materials for this purpose include cellulose sheeting, glass wool and wadding. Cellulose

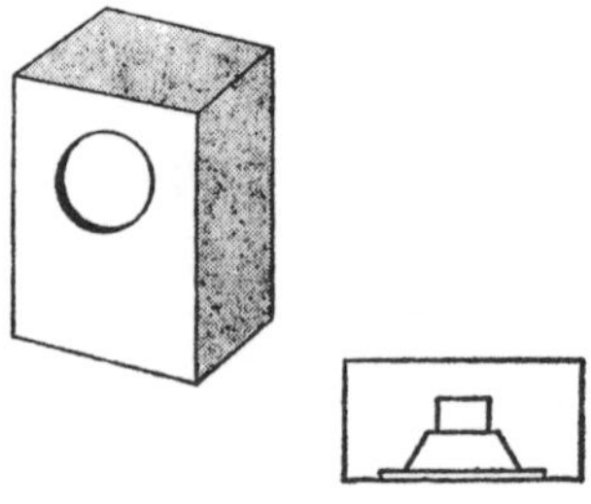

Fig. 99. Acoustic box.

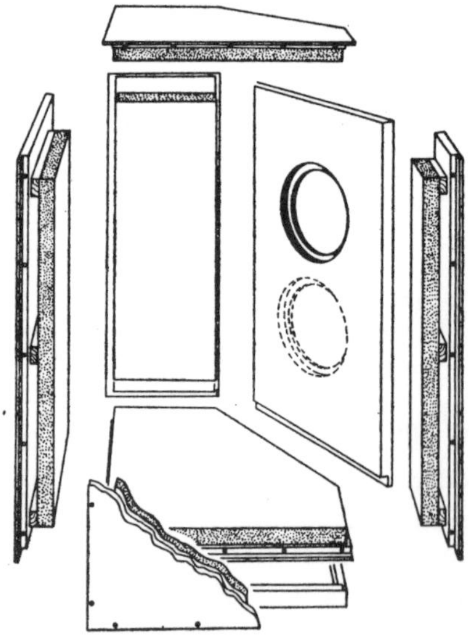

Fig. 100. Exploded view of acoustic box.

sheeting should be about 25 mm thick, and glass wool or wadding may be used in layers 40 mm thick or sandwiched between coarse hessian. Cellulose sheeting and glass wool are obtainable from builders' merchants.

The joints used in the box must fit really closely, as it is imperative that the box should not leak, and its walls must not be vibrated by the energy that will be produced inside it, so that timber about 25 mm thick should be used (for drawings of acoustic boxes, see pages 76-83).

Having looked at the shape and construction of the acoustic box, let us now examine the effect that the volume of air present inside it has on the sound-reproduction, which is by no means slight, and manifests itself in a rise in the resonant frequency of the speaker. The movements of the cone are slowed down because of the braking action of the air in the box (fig. 101), and the resultant rise in the resonant frequency is shown diagrammatically in fig. 102.

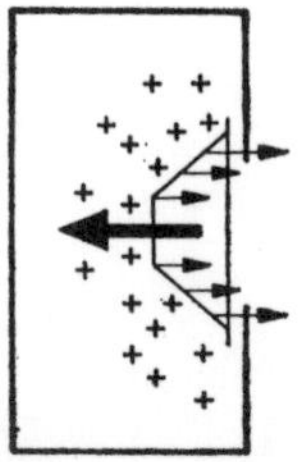

Fig. 101. In the acoustic box, the speaker undergoes the "braking" action of the air inside the box.

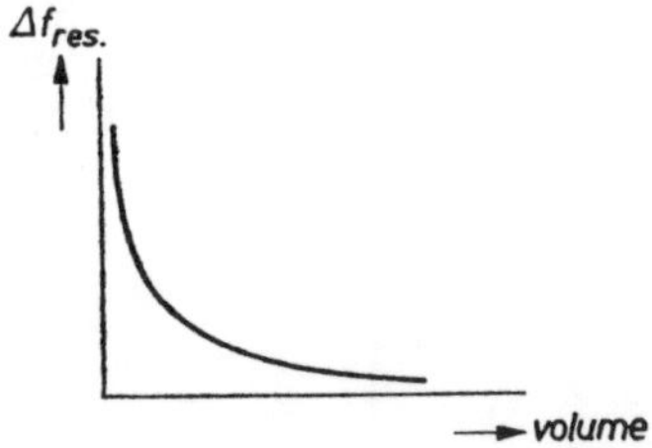


Fig. 102. Graph of the resonant frequency of the acoustic box as a function of the volume of the box.

The acoustic box may be compared to the infinite baffle, since, in both cases, the vibrations originating from the back of the cone cannot reach the front of it. There is an essential difference, however, caused by the air behind the cone, which, because of its restricted volume, can affect the cone's movements (standing wave). This volume of air is, of course, infinite in the case of the infinite baffle. The reaction of the air in the box can, however, favourably affect the attenuation of the speaker, and this is particularly important at the resonant frequency. The acoustic box is eminently suitable for quality reproduction due to its excellent baffle effect and the attenuation that has just been mentioned. The table below gives the advantages and drawbacks of this type of loudspeaker box.

*Advantages*

1. Good bass response
2. Good attenuation

*Disadvantages*

1. Construction critical
2. Rather large

**The bass-reflex cabinet**

The main difference between the bass-reflex cabinet (fig. 103) and the acoustic box lies in the vibrations emitted by the rear of the cone. In the former this energy is not absorbed, as it is in the acoustic box, but used to amplify the energy put out by the front of the cone. This can take place only when a "+" from the rear of the cone appears a short time after its corresponding "—" at the front of the speaker. This delay is also known as a phase-shift, and is produced in the bass-reflex cabinet by a cylindrical aperture in the front of the box (fig. 104). It is difficult to explain how this produces phase-shifting, but it can, perhaps, be illustrated by the analogy that exists between mechanical and electrical values and quantities. The volume of air in the box

Fig. 103. bass-reflex cabinet.

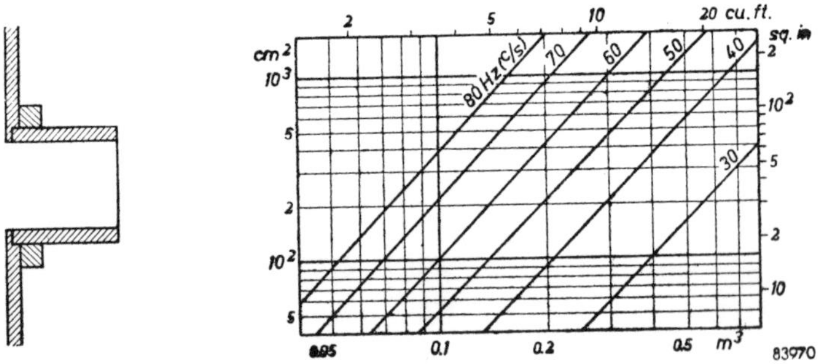


*a* *b*

Fig. 104*a*. Tunnel of Bass-reflex cabinet.
Fig. 104*b*. Volume and aperture area of bass-reflex cabinet at various frequencies.

may be compared to a capacitor, and the mass of air in the cylinder formed by the aperture to a self-inductance or coil. By judicious selection and matching of the volume of air in the box and the size of the aperture, the resonant frequency of the box can be tuned to that of the speaker inside it, although the latter will tend to alter in time, thus upsetting the balance.

The sound-intensity produced by a bass-reflex cabinet is greater than that of an acoustic box, but the resonant frequency of the speaker is raised considerably by the column of air in the box, preventing the very lowest notes from being reproduced. Unlike the acoustic box, the resonant frequency of the speaker is not attenuated, but rather amplified, so that, when better quality reproduction is required, the acoustic box is to be preferred.

The interior of a bass-reflex cabinet should be lined with sound-damping material, the position of which is clearly shown by the partly exploded view of a cabinet in fig. 105. The materials used and methods of applying them have already been discussed in connection with the acoustic box, and the cabinet itself should be made of solid material at least 16 mm thick.

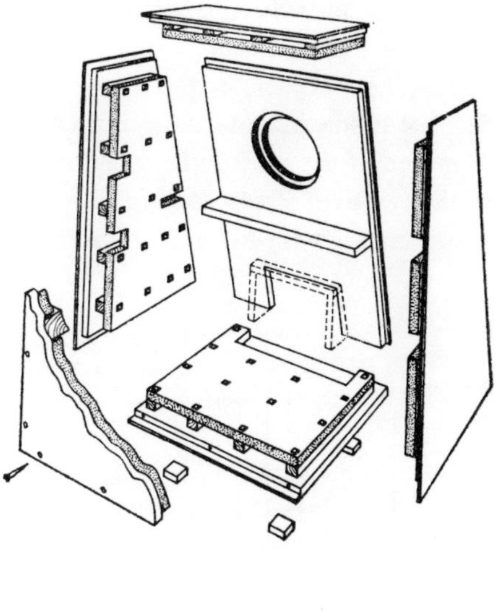

Fig. 105. Exploded view of bass-reflex cabinet.

Drawings 1, 2 and 3 in Chapter 9 show the front panel mounted at an angle of 75° to the base, to ensure good treble response (see fig. 106). The equivalents of these drawings are published under the numbers 4, 5 and 6, but in these cases the angle between front panel and base is 90°.

There are other types of loudspeaker box besides the two mentioned above,

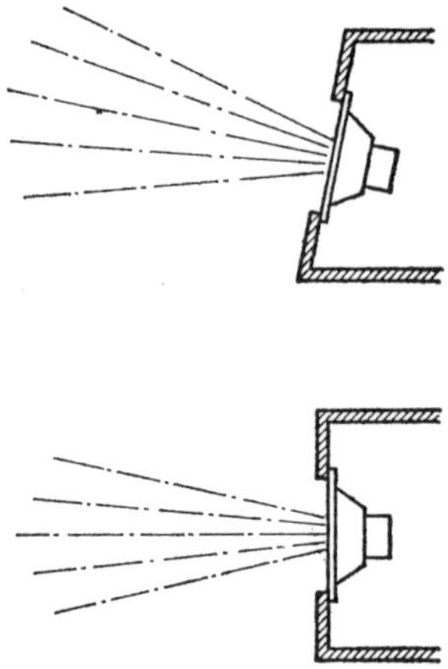

Fig. 106. Effect of the slope of the front of the box on the direction in which the treble notes are radiated.

but they are rather difficult to make and are so large that there would be little point in including them in this book. Nevertheless, there is one type that does bear mention, the "open-backed" type, although this is, to some extent, a misnomer, as the back is not really open, but covered off with a sheet of perforated metal, in turn covered with cloth on both sides (fig. 107). The area of the perforated plate and the size of the perforations are suited to the resonant frequency of the speaker. This box, which is very simple to make, gives good results. Metal sheeting with standard 0.25 $cm^2$ perforations is readily obtainable.

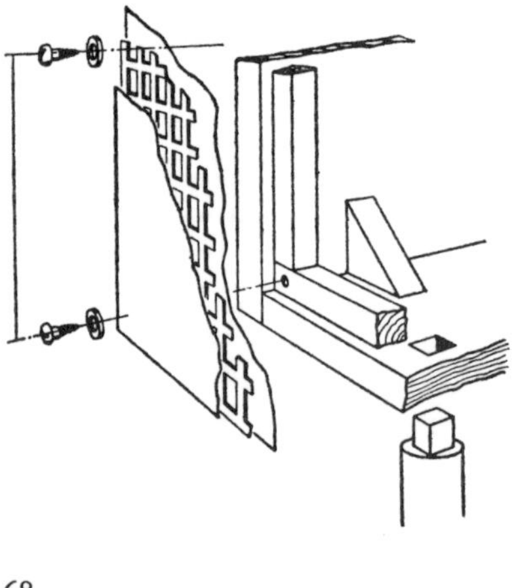

Fig. 107. Rear wall of speaker box.

CHAPTER 8

# BUILDING LOUDSPEAKER BOXES ONESELF

This chapter deals with a few of the timber-joints useful in the construction of speaker boxes, and also with some other facets of building them such as the design of the baffle aperture, and finishing the box.

## Timber-joints

The most common method of joining two pieces of wood together at right angles is shown in fig. 108, and is the simple butt joint. The disadvantage of this type, however, is the fairly small area of the contacting surfaces; the sawn-off end of piece *A* is never, in point of fact, completely smooth and square. Fig. 109 shows a number of joints often used with great success, although most of them need special tools to make them fit really accurately,

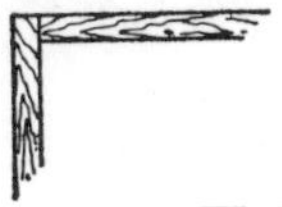

Fig. 108. Butt joint.

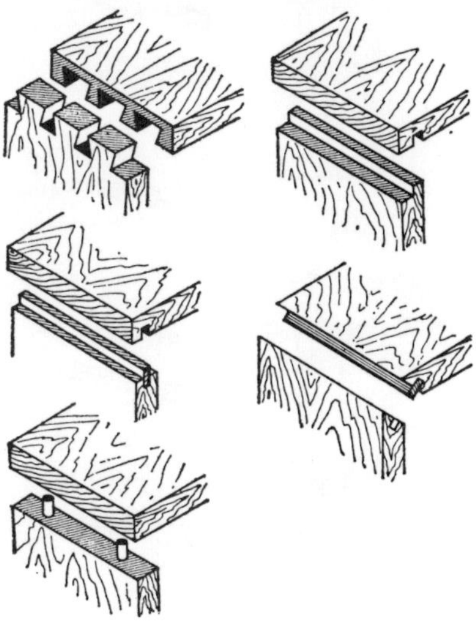

Fig. 109. Various timber-joints.

and to make the construction of these boxes as simple as possible, only the most elementary kinds of joint will be dealt with here.

## Single rabbet

Fig. 110 shows a single-rabbetted joint, which is very easy to make with the help of an adjustable-blade circular saw, shown in fig. 111. If none is available, the other method of making it is equally good but takes longer, and is

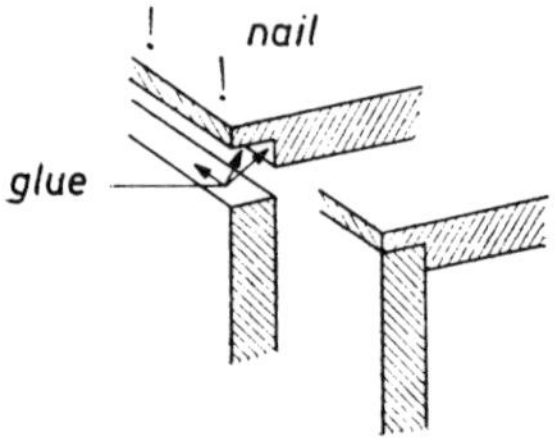


Fig. 110. Single-rabbet joint.

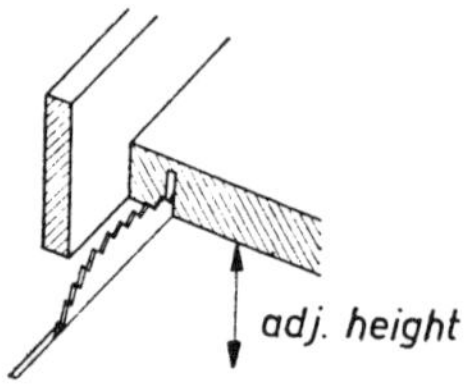


Fig. 111. Sawing a single rabbet with an adjustable-blade circular saw.

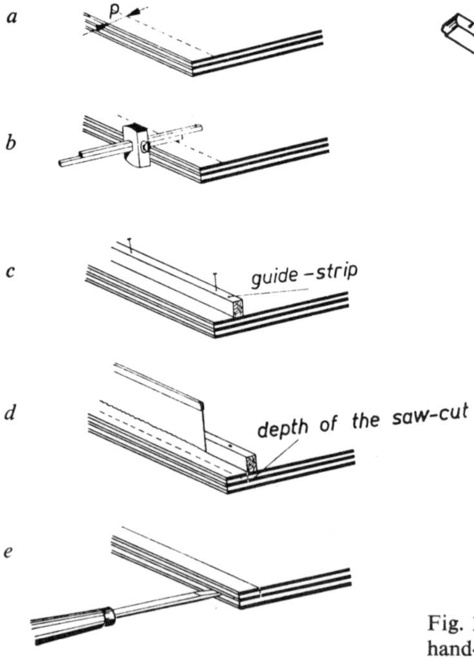


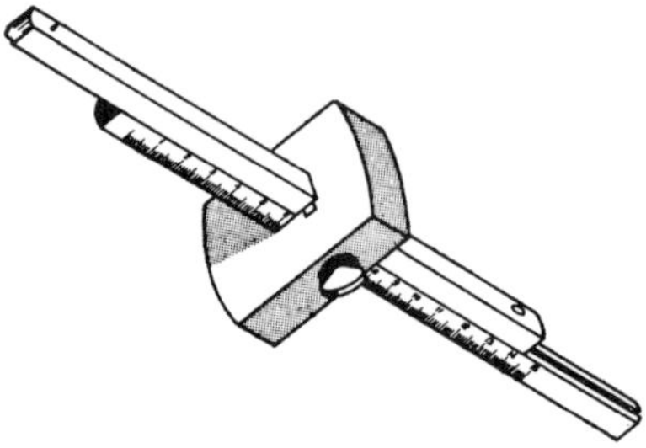

Fig. 113. The marking gauge.

Fig. 112. Sawing a single rabbet with a hand-saw.

illustrated in figs. 112*a* to *e* inclusive. Fig. 112*a* shows the piece of timber in which the rabbet is to be cut, marked out with a marking gauge, *p* representing the depth of the rabbet. The gauge is shown in fig. 113, and the method of using it in fig. 112*b*. A strip of wood 20 mm high is then tacked along the score mark made by the gauge, as shown in fig. 112*c*, and the first cut may then be made with a tenon saw (fig. 112*d*). The depth of the saw-cut must, of course, be as uniform as possible over the whole length of the timber, although, with this method, it will never be exactly the same depth all the way along. If plywood is used, and this is recommended for all the boxes dealt with in the following chapter, the saw-cut can end at one of the plies, whereupon the waste wood can be chiselled out (fig. 112*e*).

## Double rabbet

Double-rabbetted joints are also frequently used (fig. 114). The advantage of this type of joint is not only that the area of contact is increased, but also that it resists distortion. The usual thickness of the timber in the first rabbet *a* is about 3 mm. The contacting surfaces I and II must be a good fit, although there is a certain tolerance on surface III.

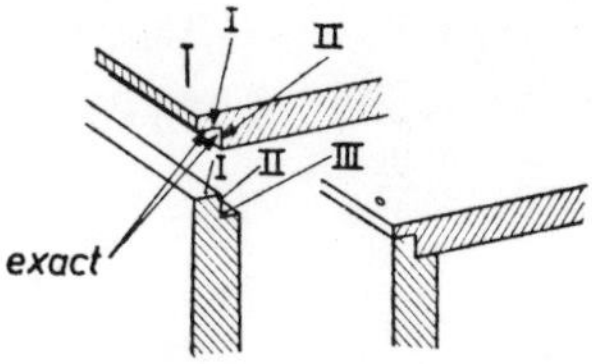


Fig. 114. Double-rabbet joint.

## A few non-right-angled joints

Both the abovementioned joints are used to join pieces of timber together at right angles. Where an angle different from this is required, however, the edges of the timber have to be bevelled (fig. 115), which can be done with the aid of the circular saw by raising the transverse guides to the correct height (fig. 116). If no such saw is available, however, and the job has to be done by hand, it is best to use a plane and make up a checking template from two strips of wood joined together at the appropriate angle (fig. 117). This is the type of joint used in the acoustic box.

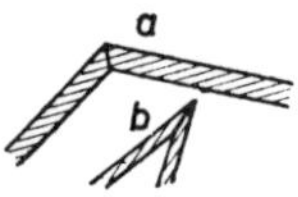


Fig. 115. Joints at more (*a*) and less (*b*) than 90°.

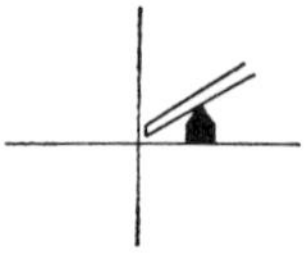

Fig. 116. Method of bevelling using a circular saw.

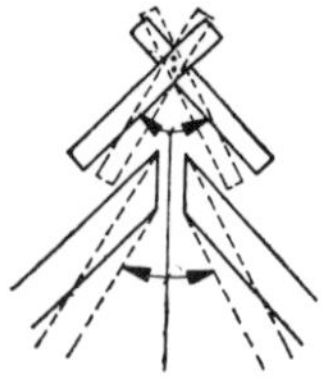

Fig. 117. Template for checking angle of bevel.

Bass-reflex cabinets with sloping fronts need another kind of non-right-angled joint, the shape of which is shown in the drawings in chapter 9 (fig. 118). It will be noted that one bevel is made in this case, which makes a very good joint. Because the design of this type is simple, it is recommended that the method shown in fig. 119 be followed.

Fig. 118. Angled joint between front and top of bass-reflex cabinet.

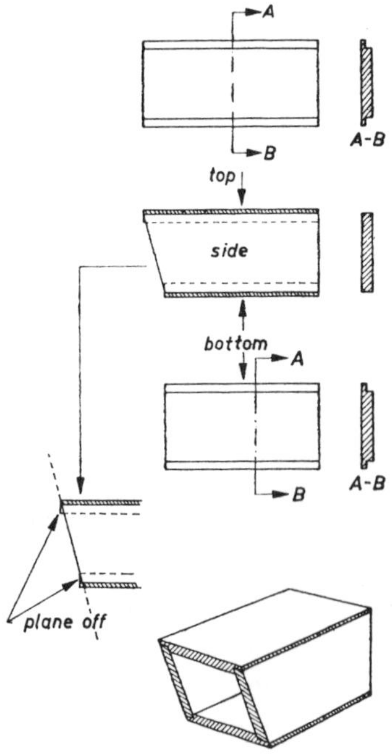


Fig. 119. Simple method of making the joint shown in fig. 118.

**Fixing the joint**

Once the joints have been shaped, the various parts have to be fixed together, and for this it is best to use a cold-water glue, with a few pins inserted here and there along the joint to reinforce it. To give the box as good a finish as possible, headless pins should be used.

**The baffle aperture**

The simplest way of making the baffle aperture is to mark out a circle with a pair of compasses, drill a hole just inside the circumference of the circle, and finally to cut out the waste wood using a fretsaw, or scroll saw for very small baffles (fig. 120).

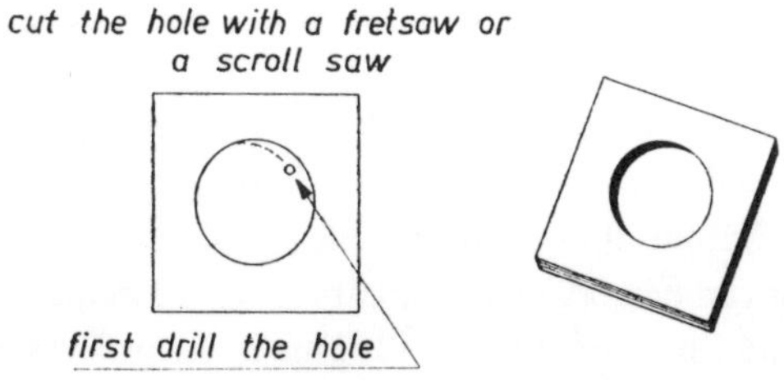


Fig. 120. Making the baffle aperture.

**Feet**

Boxes designed to stand on the floor should be provided with feet, fig. 121 showing a few ways of doing this.

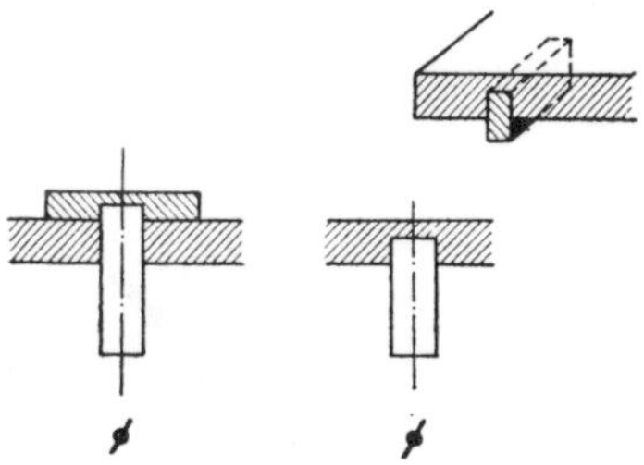

Fig. 121. Various methods of securing feet to the base of the box.

### The kind of timber to use

Plywood is the most suitable wood for making loudspeaker boxes because of its mechanical strength and the ease with which it can be worked, block-board, in which the central ply consists of strips of timber (fig. 122) being really too brittle for this purpose.

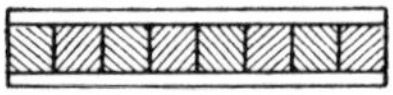

Fig. 122. Section through blockboard.

### Finishing

There is a number of ways of giving the cabinet a good finish, including veneering, varnishing, painting and staining. Only veneering will be explained in detail here, since the other methods are supposed to well-known.

Veneer, a thin layer of wood, is applied to the surface of the cabinet and held by some kind of adhesive, of which there are several types on the market. If the adhesive used is in the form of a solution, both the veneer and the cabinet are coated with it and allowed to dry off for about twenty minutes, whereupon the veneer can be applied to the cabinet in the manner shown in fig. 123 and smoothed and pressed down well. This latter process may be done in a number of ways, fig. 124 showing the method of using a block hammered down over the entire surface, while the use of the veneering hammer is illustrated in fig. 125. After the glue has been allowed to dry completely, the projecting edges of the veneer must be cut off with a veneering saw or a stout, sharp knife, the last remaining slight projections being then glasspapered off.

The cabinet may be brought to a high gloss by alternately glasspapering

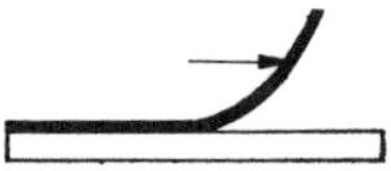

Fig. 123. Applying veneer.

Fig. 124. Tapping the veneer down.

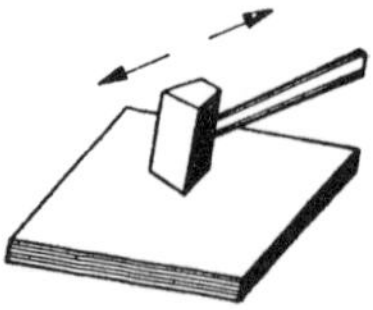

Fig. 125. Using the veneering hammer.

and applying clear lacquer, although this process will have to be repeated at least ten times before a really brilliant polish is obtained.

When the actual timber of the box has been properly finished, the appearance of the cabinet can be even further enhanced by the application of trim-strips and by covering the front with loudspeaker fabric, which is a kind of cloth having good acoustic properties. It is infinitely preferable to any other kind of fabric, however well its colour and weave might suit the purpose, although a very open-weave material could be used at a pinch, provided that it does not affect the treble response.

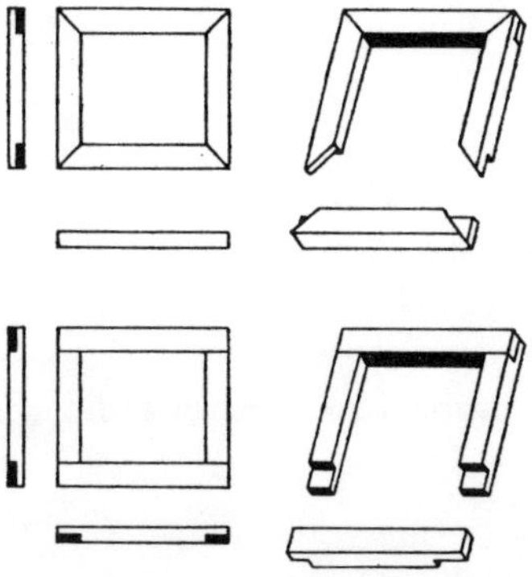

Fig. 126. Some types of frame.

CHAPTER 9

# DRAWINGS OF CABINETS

The following pages contain a large number of drawings of cabinets, offering a wide selection to suit any installation. Each drawing carries a symbol, the key to which is given below:

*Z* = excellent response
*H* = good response
*S* = suitable for stereo-installations
*M* = suitable for mono-installations
*B* = suitable for building into a book-case
*HT* = primarily for treble response
*E* = suitable as an additional speaker for radio sets, tape recorders, etc.

The types are divided up as follows:
seven drawings of acoustic boxes,
eight drawings of bass-reflex cabinets, and
nine drawings of boxes based on different principles from the above.

*Acoustic box 1*

Number of loudspeakers: 2
Loudspeaker types: 9710, 9710 M or 9710 AM
*Z, S, M*

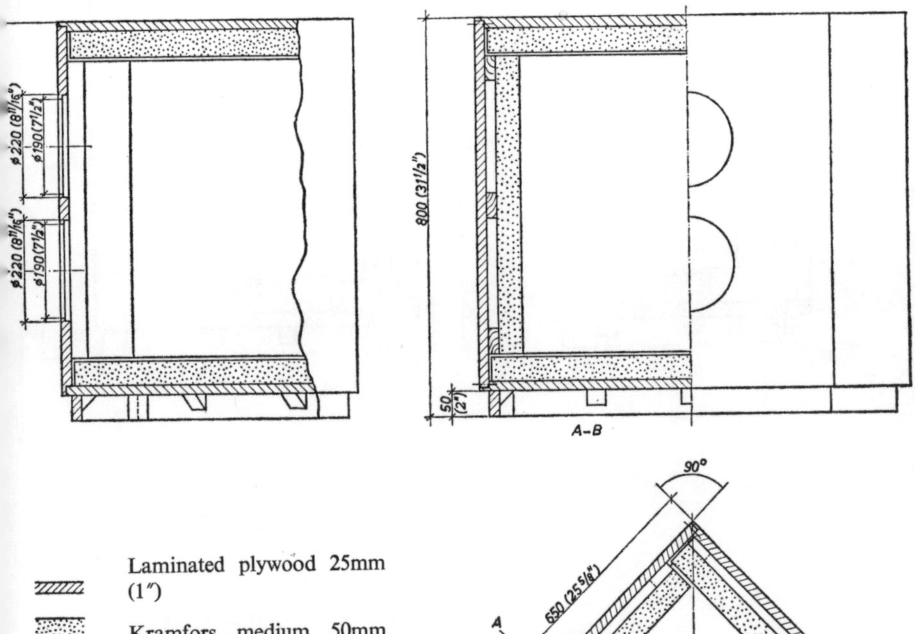


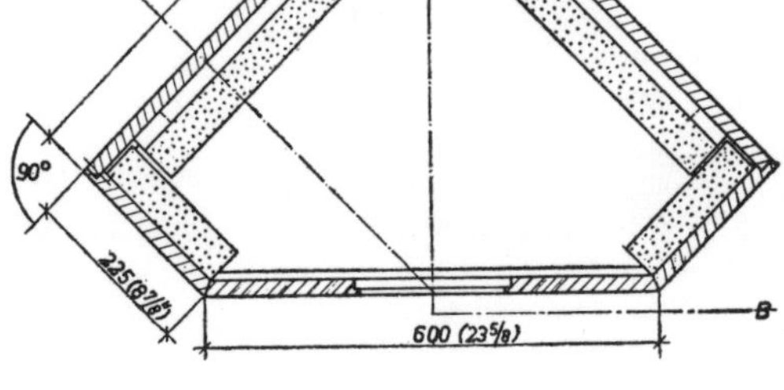


Laminated plywood 25mm (1″)

Kramfors medium 50mm (2″) or glasswool blanket

*Acoustic box 2*

Number of loudspeakers: 1

Loudspeaker types: 9710, 9710 M or 9710 AM

*Z, S, M*

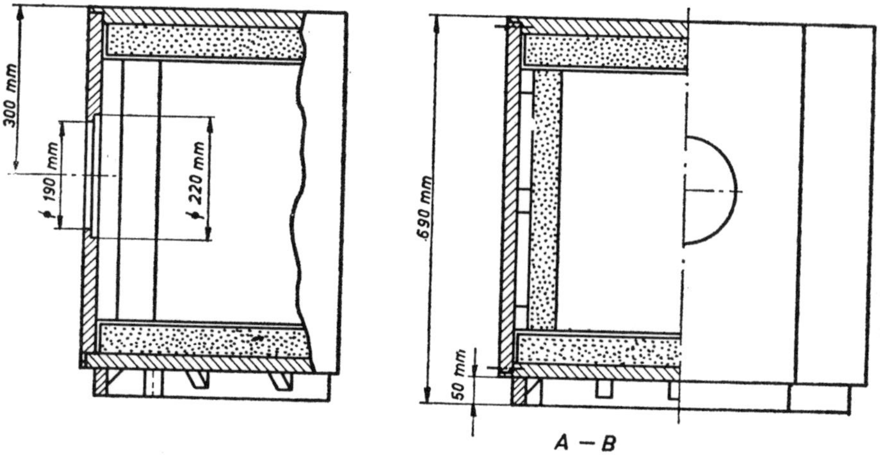


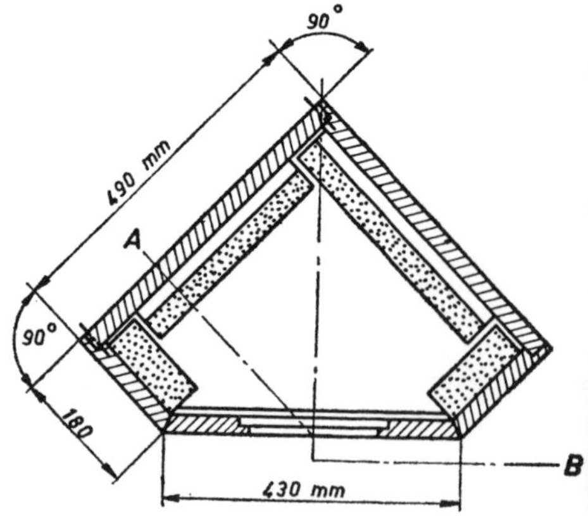

*Acoustic box 3*

Number of loudspeakers: 1
Loudspeaker types: AD 4000, AD 4000 M or AD 4000 AM

*Z, S, M*

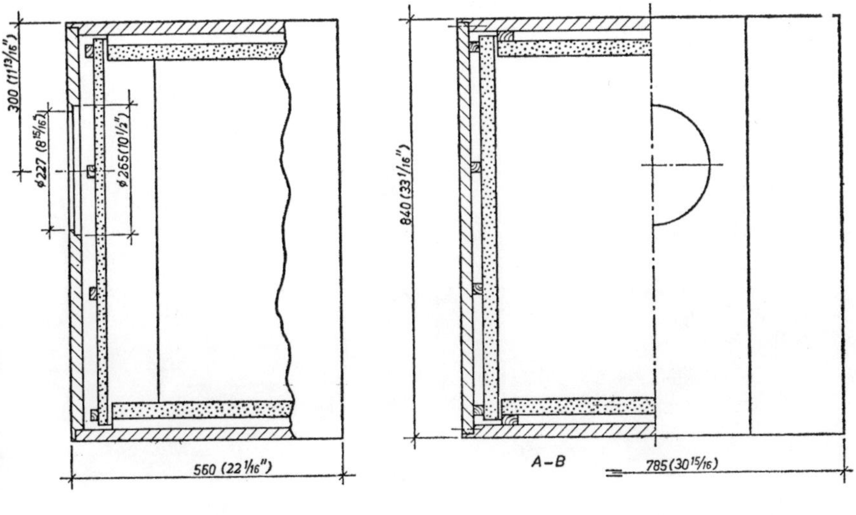


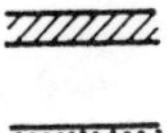

Laminated plywood 25mm (1″)

Kramfors medium 50mm (2″) or glasswool blanket

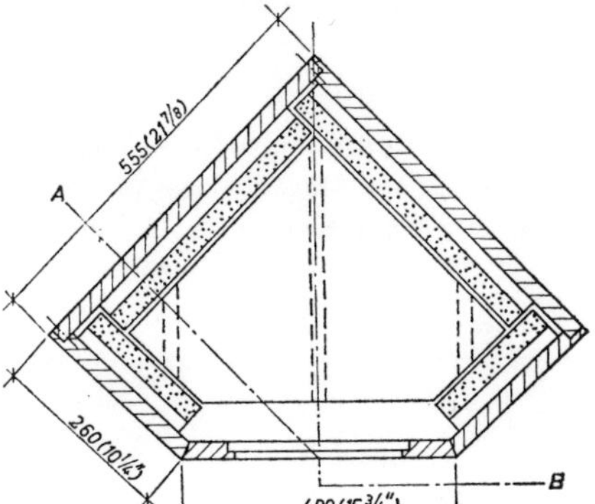

*Acoustic box 4*

Number of loudspeakers: 1

Loudspeaker types: AD 4200, AD 4200 M or AD 4200 AM;
AD 5200, AD 5200 M or AD 5200 AM

*Z, S, M*

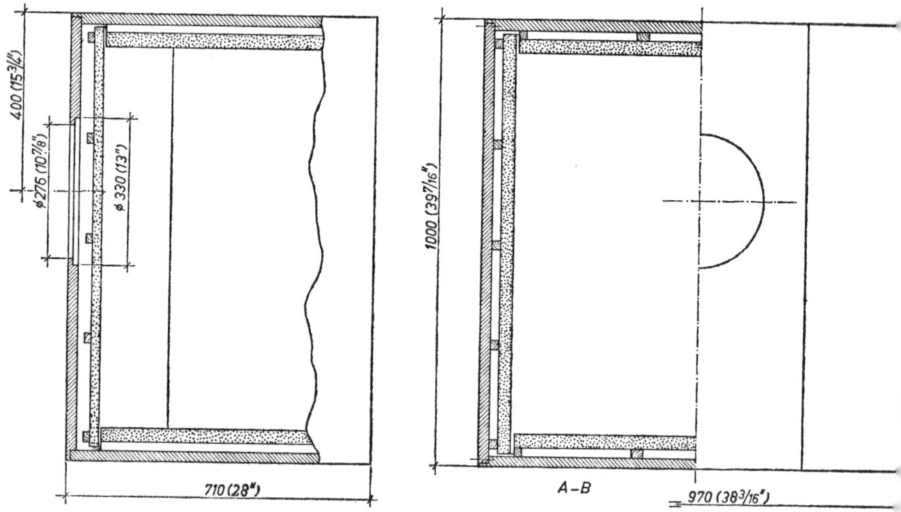


Laminated plywood 25mm (1″)

Kramfors medium 50mm (2″) or glasswool blanket

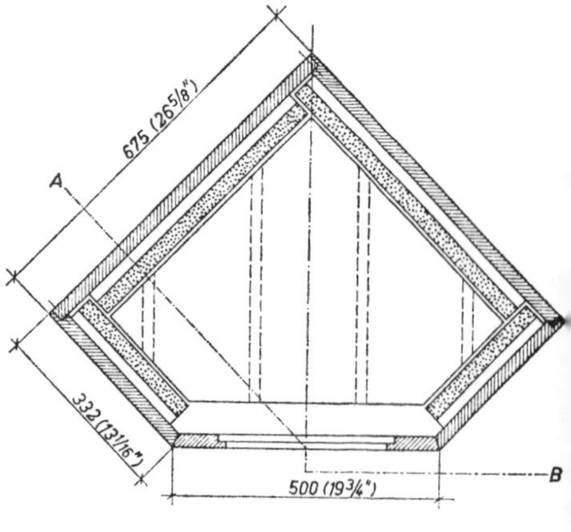

## *Acoustic box 5*

Number of loudspeakers: 1
Loudspeaker types: 9710 M or 9710 AM

*Z, M*

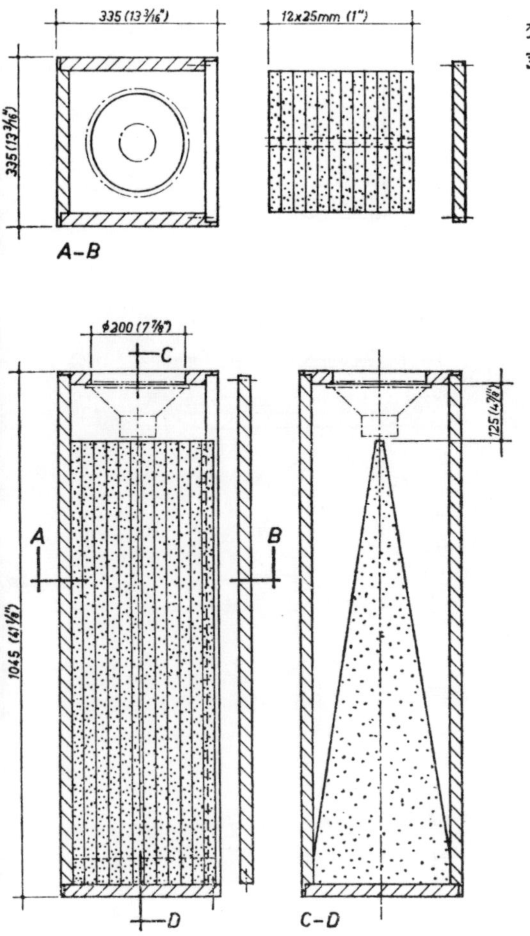


Laminated plywood 25mm (1″)

Kramfors medium 50mm (2″) or glass-wool blanket

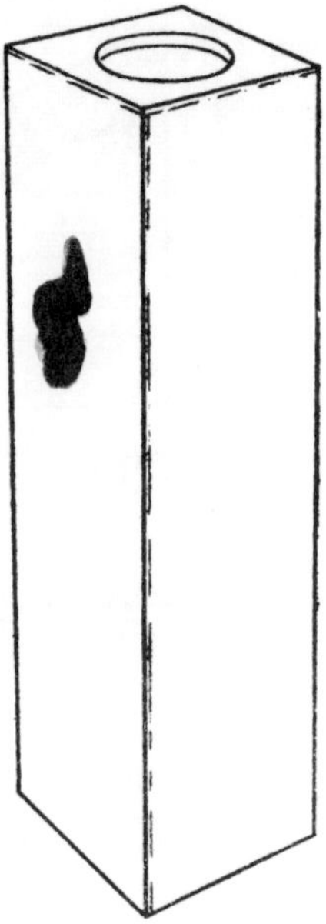

*Acoustic box 6*

Number of loudspeakers: 1

Loudspeaker types: 9710 M or 9710 AM

*Z, S, M, B, HT*

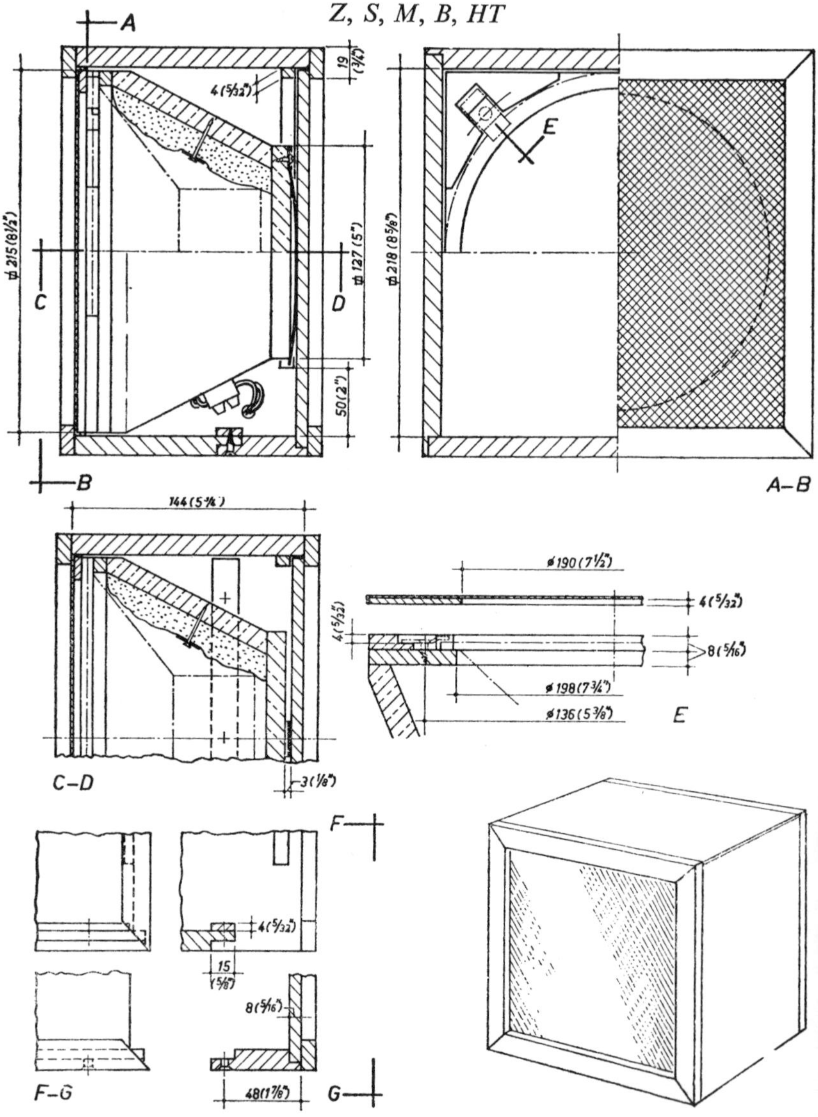


Laminated plywood 12mm

Chipboard 12mm

Kramfors medium 25mm (1″) or glasswool blanket

## *Acoustic box 7*

Number of loudspeakers: 1
Loudspeaker types: 9710 M or 9710 AM

*Z, S, M*

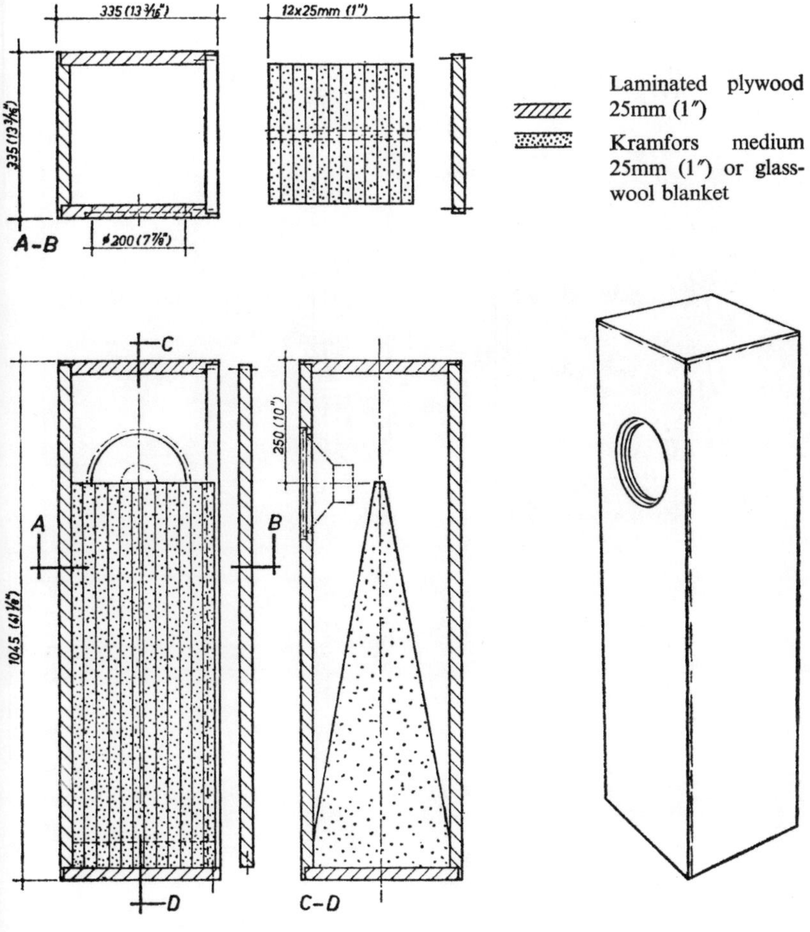

*Bass-reflex cabinet 1*

Number of loudspeakers: 1

Loudspeaker types: 9710, 9710 M or 9710 AM

*H, S, M*

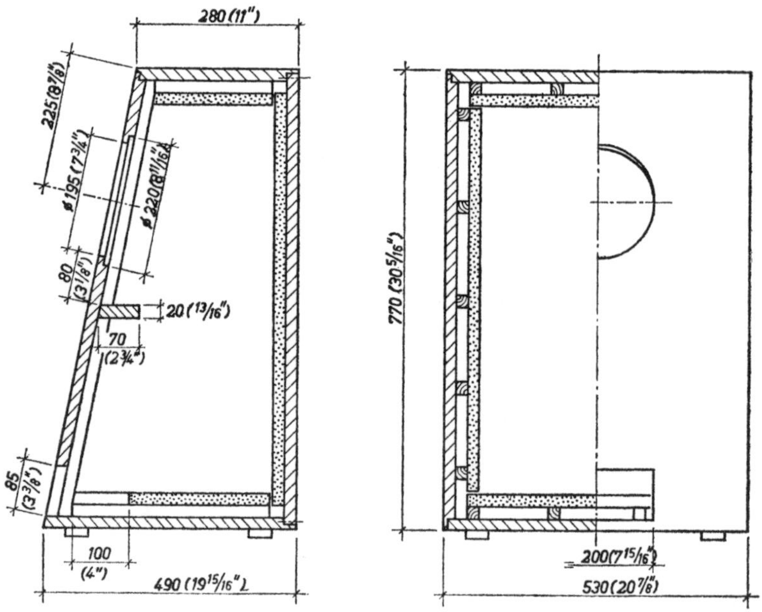


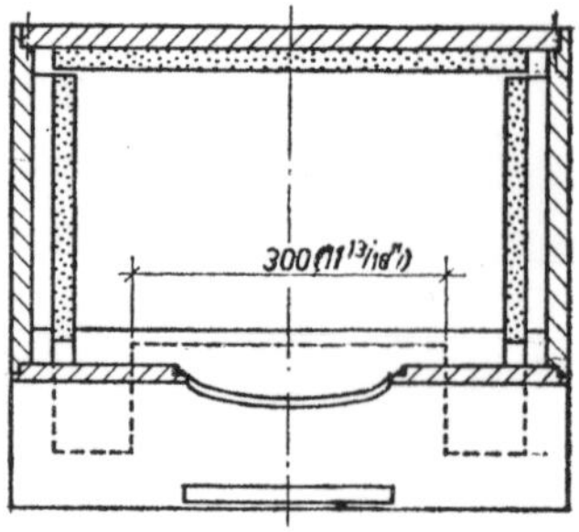


Laminated plywood 20mm

Kramfors medium 25mm or glasswool blanket

*Bass-reflex cabinet 2*

Number of loudspeakers: 1

Loudspeaker types: AD 4000, AD 4000 M or AD 4000 AM

*H, S, M*

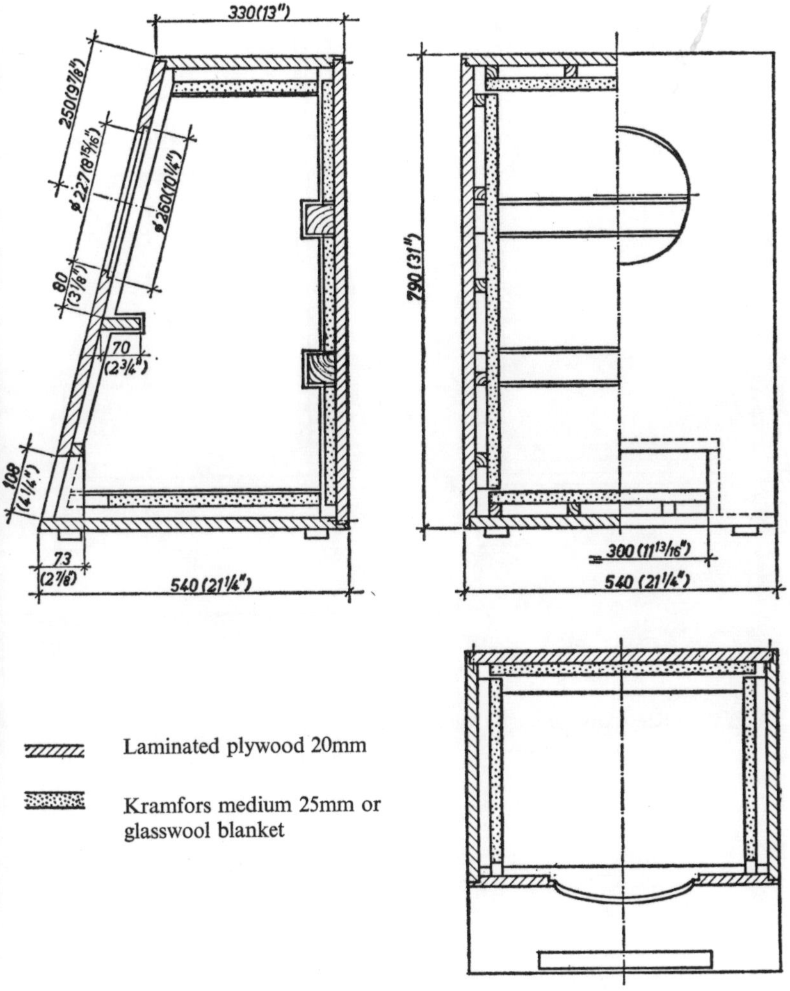

*Bass-reflex cabinet 3*

Number of loudspeakers: 1

Loudspeaker types: AD 4200, AD 4200 M, AD 4200 AM
AD 5200, AD 5200 M or AD 5200 AM
*H, S, M*

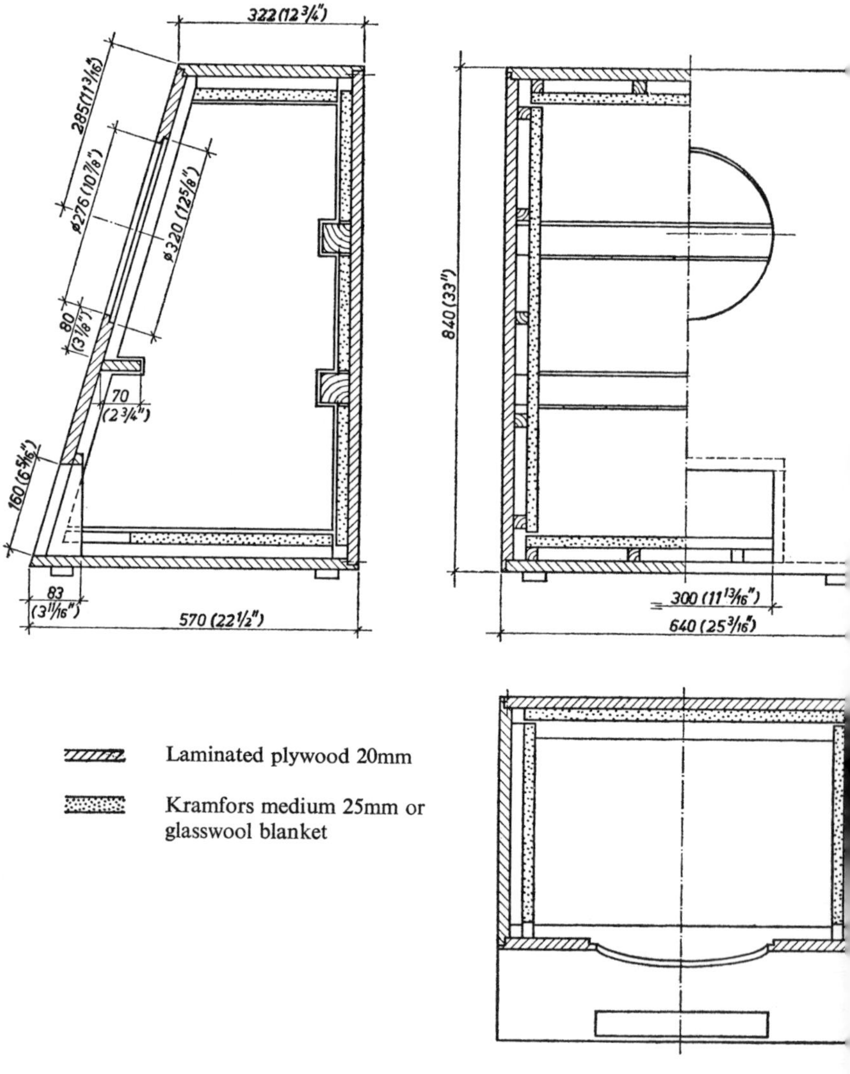


Laminated plywood 20mm

Kramfors medium 25mm or glasswool blanket

*Bass-reflex cabinet 4*

See bass-reflex cabin

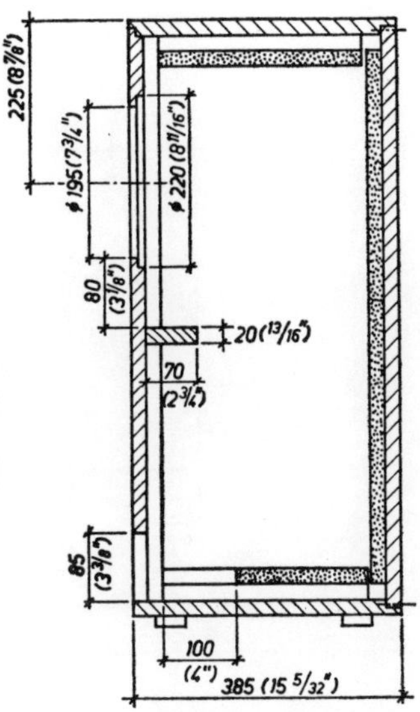


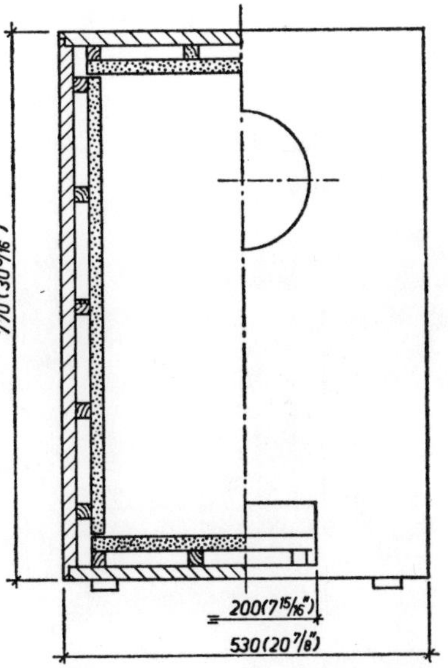


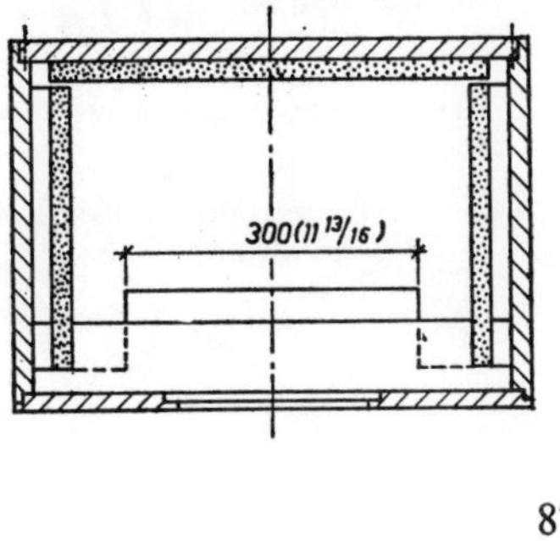


Laminated plywood 20mm

Kramfors medium 25mm or glasswool blanket

## *Bass-reflex cabinet 5*

See bass-reflex cabinet 2

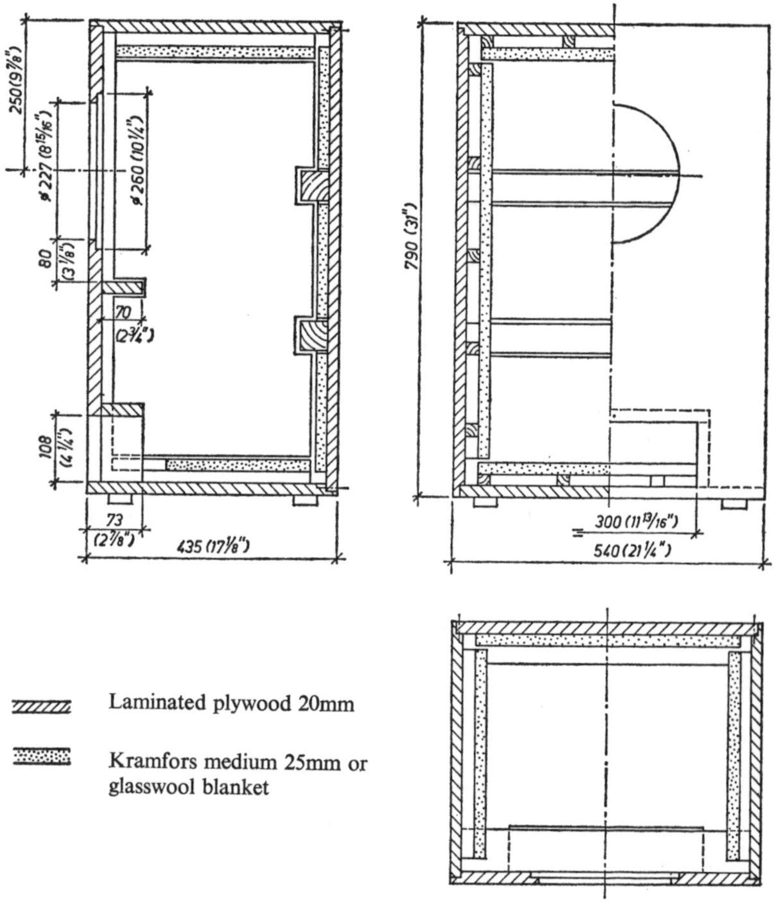

## *Bass-reflex cabinet 6*

See bass-reflex cabinet 3

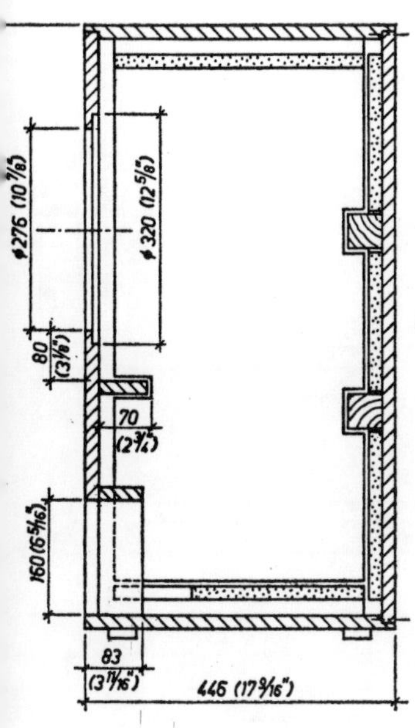


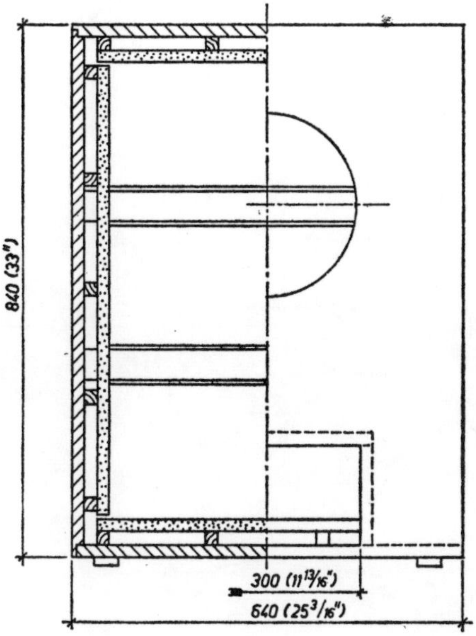


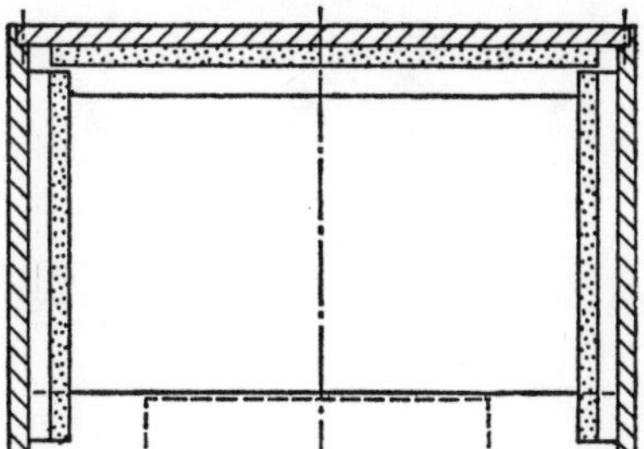

Laminated plywood 20mm

Kramfors medium 25mm or glasswool blanket

*Bass-reflex cabinet 7*

Number of loudspeakers: 1

Loudspeaker types: AD 3800/06, AD 3800/06 M, or AD 3800/06 AM

*H, S, M*

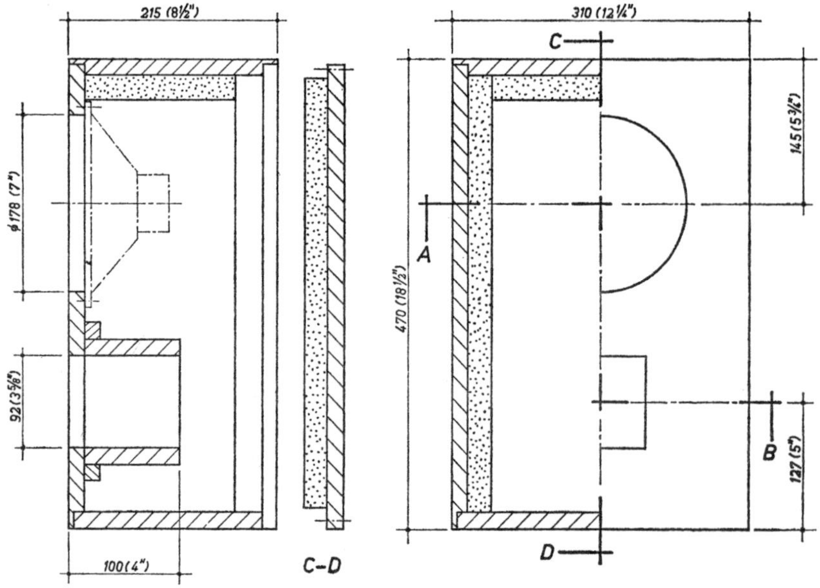


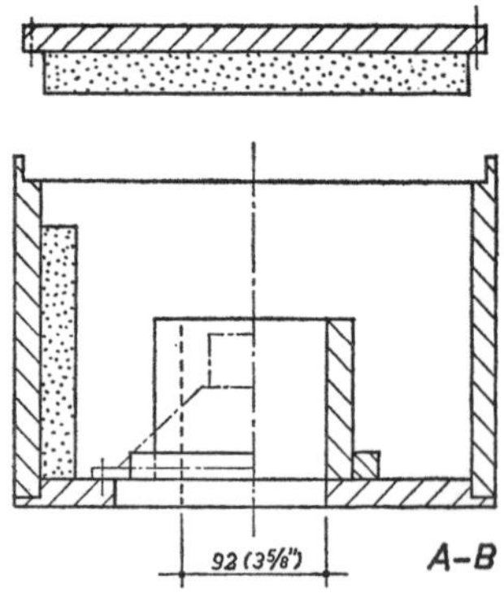


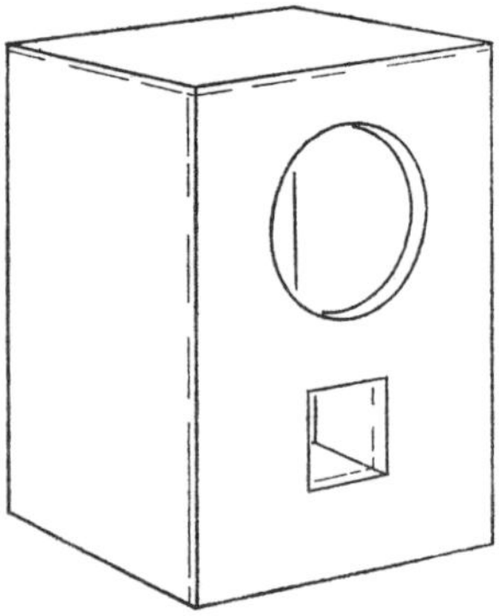

▨ Laminated plywood 15mm

░ Kramfors medium (50mm) or glasswool blanket (50mm)

*Bass-reflex cabinet 8*

Number of loudspeakers: 1

Loudspeaker types: AD 3800/06 M, AD 3800/06 AM

*H, S, M*

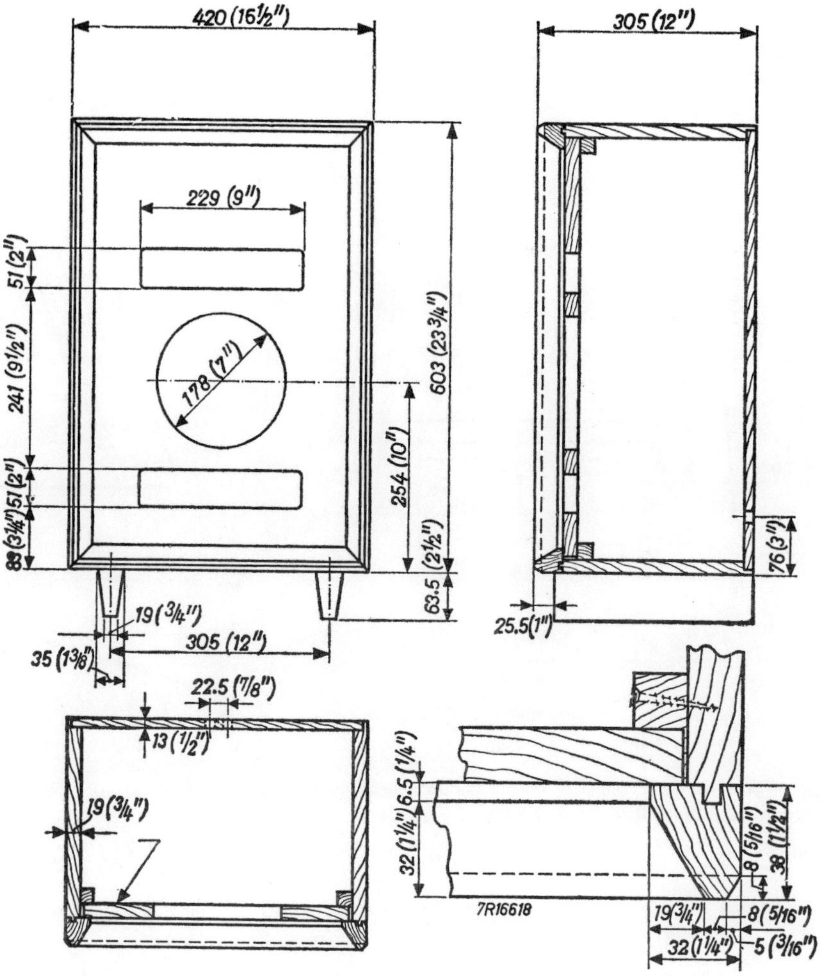

*Model 1*

Number of loudspeakers: 1

Speaker types: AD 4200, AD 4200 M, AD 4200 AM
AD 5200, AD 5200 M, AD 5200 AM
*Z, S, M*

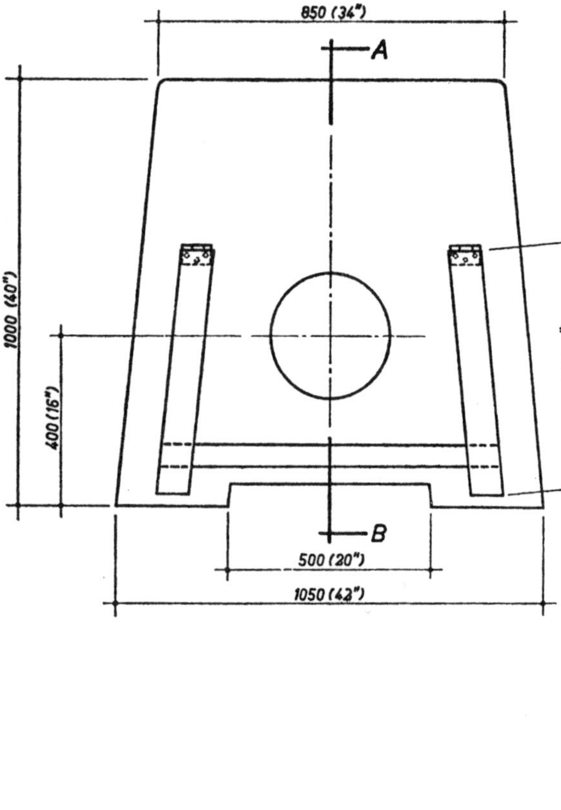


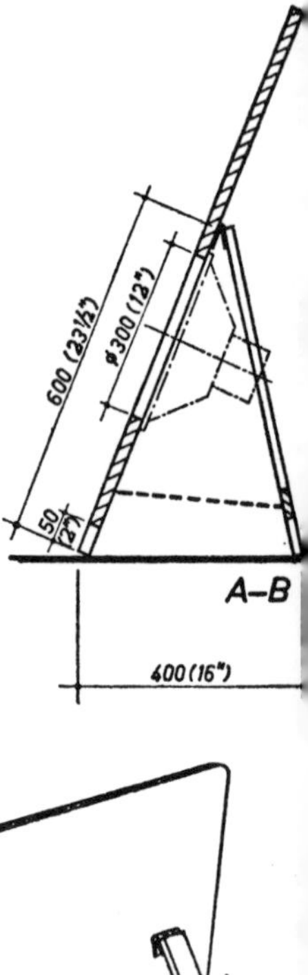

Laminated plywood 19mm

*Model 2*

Number of loudspeakers: 1

Speaker types: AD 3800/06 M, AD 3800/06 AM or AD 3800/06 *H, S, M, B, E*

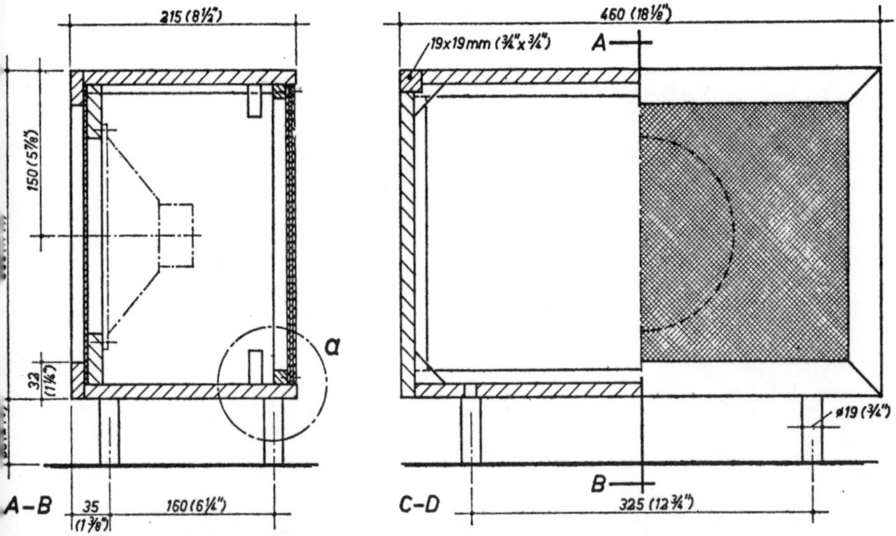


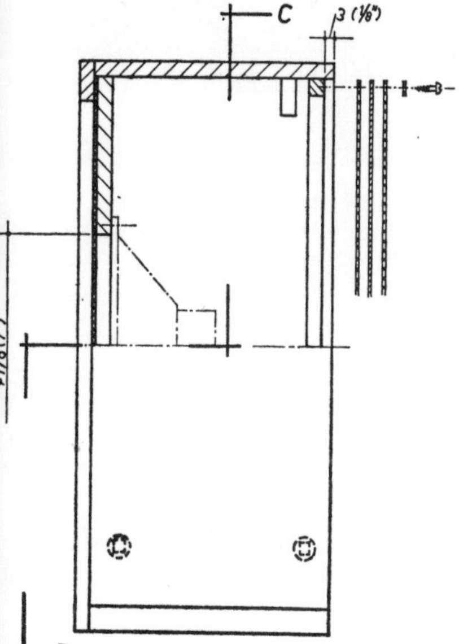


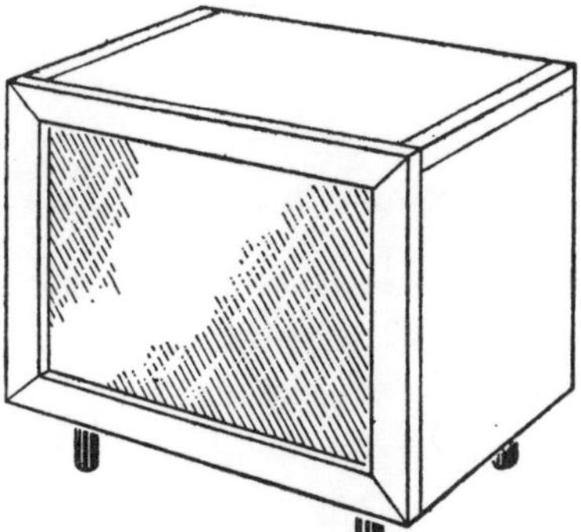

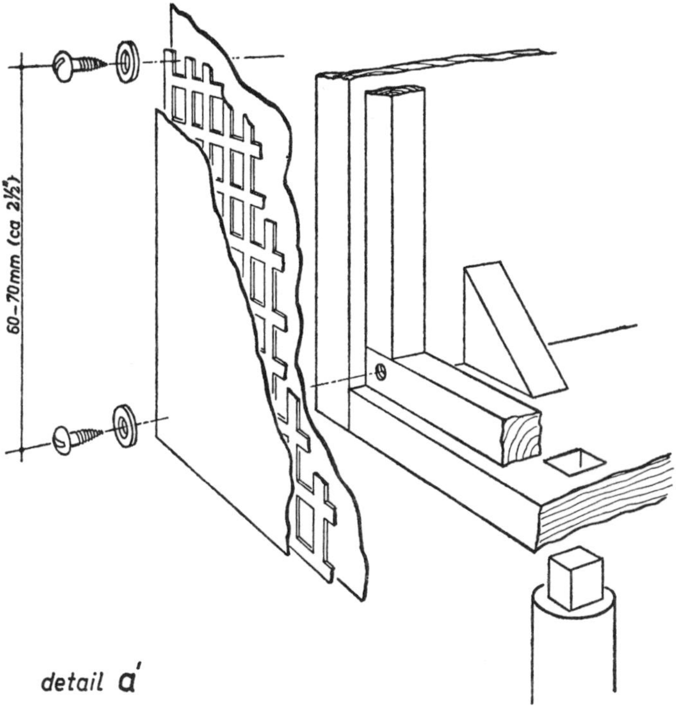


*detail a'*

Rear panel for models 2, 3 and 4 with perforated metal plate between two sheets of flannel.
The perforations are square, 0.25 $cm^2$ in area and 0.5 cm (2/10") apart.

*Model 3*

Number of loudspeakers: 1

Speaker types: 9710, 9710 M or 9719 AM

*Z, S, M, B*

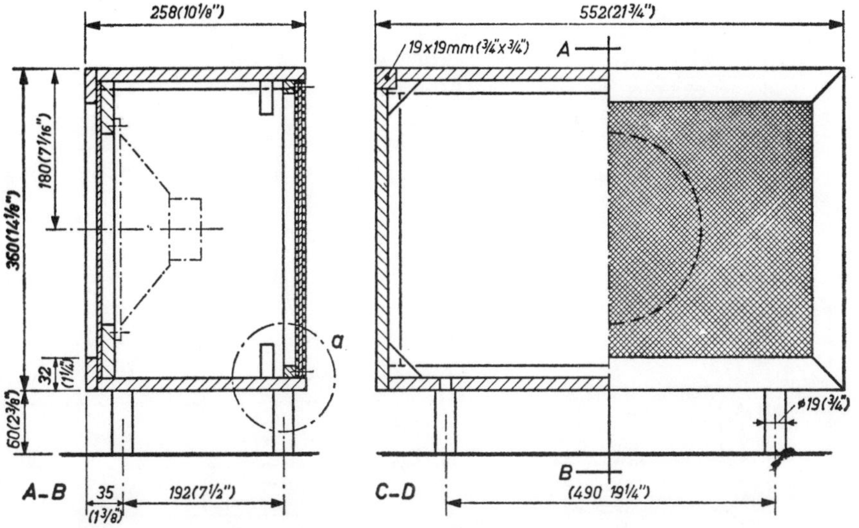


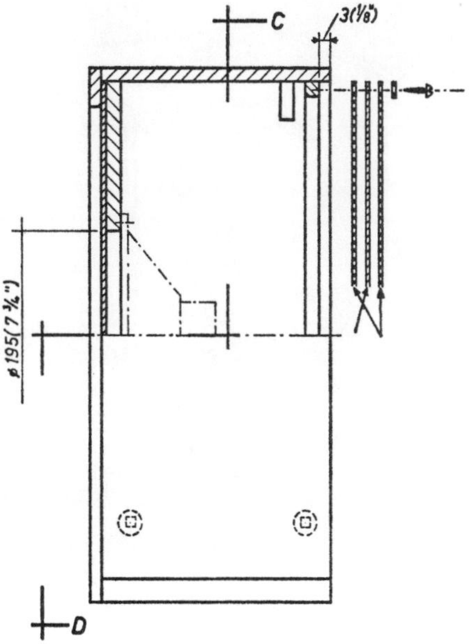


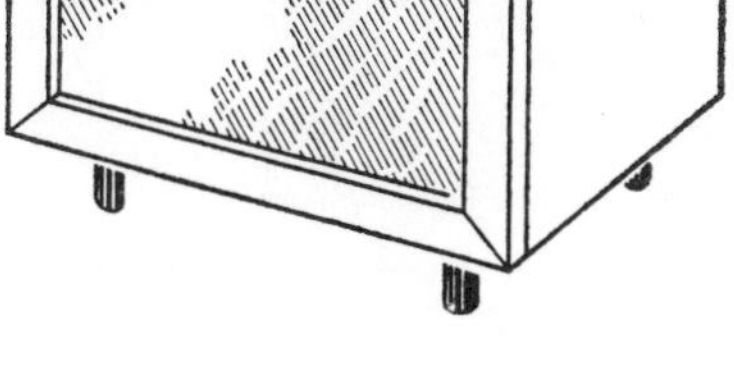

*Model 4*

Number of loudspeakers: 1

Speaker types: AD 4200, AD 4200 M, AD 4200 AM
AD 5200, AD 5200 M, AD 5200 AM
*Z, S, M*

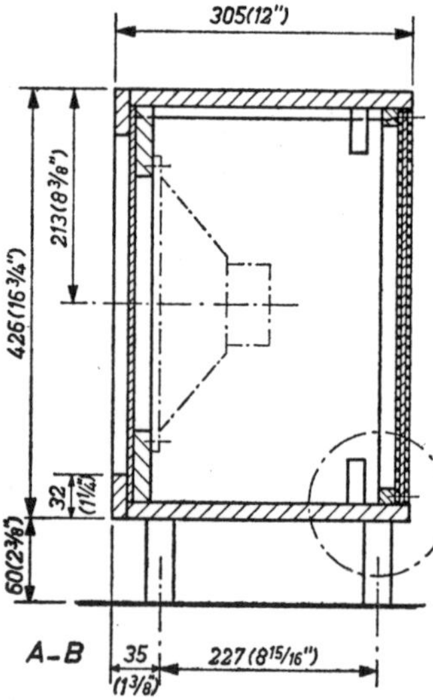


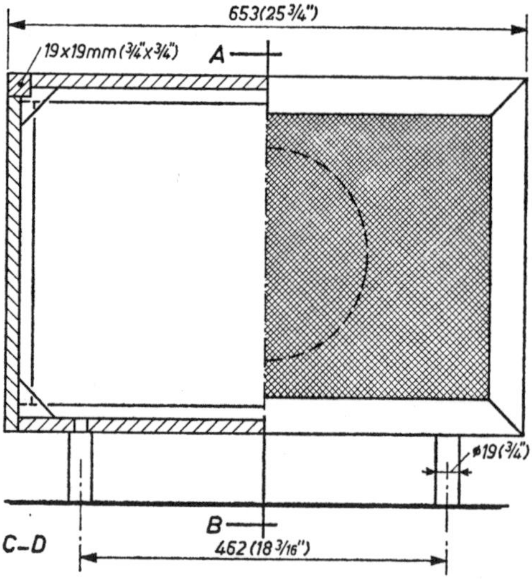


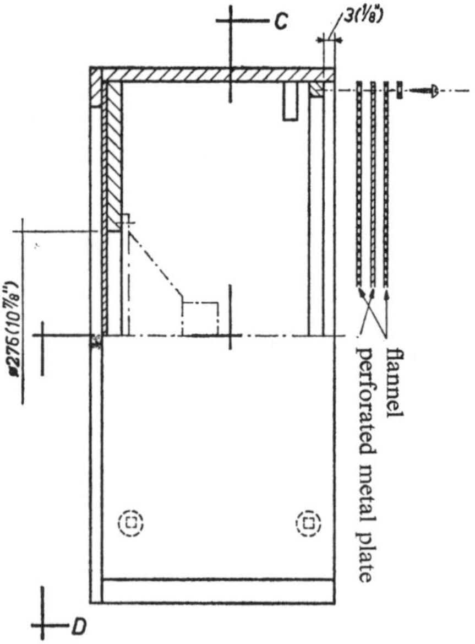


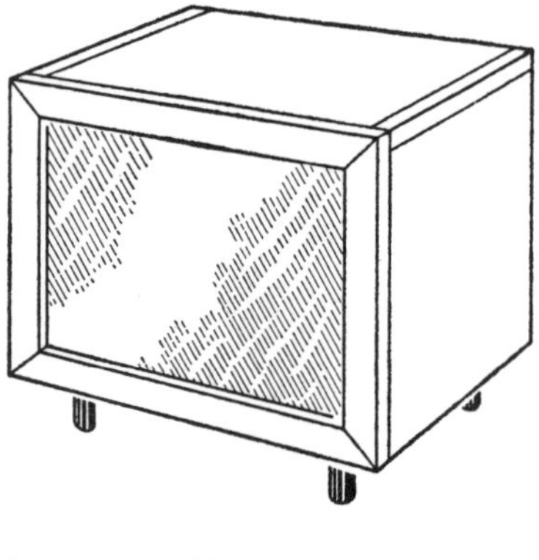

The speakers AD 4000, AD 4000 M and AD 4000 AM may also be used, but the baffle aperture has to be reduced from 276 to 227 mm.

*Model 5*

Number of loudspeakers: 1

Speaker types: AD 3800/06 M, AD 3800/06 AM

*B, HT, E*

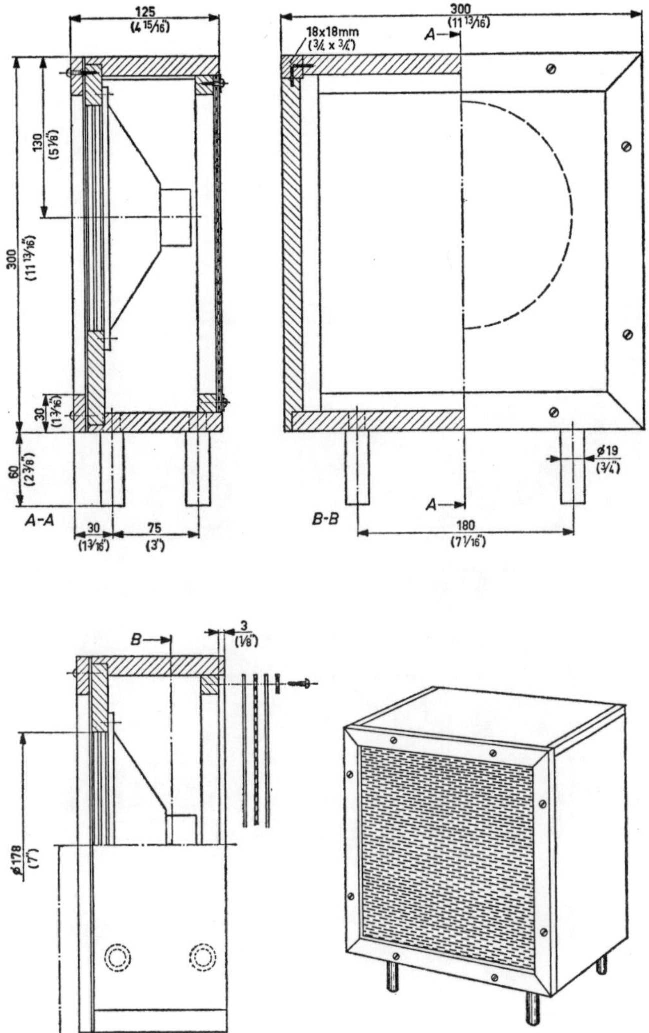

## *Model 6*

Number of loudspeakers: 2
Speaker types: 9710, 9710 M or 9710 AM
*Z, S, M, B*

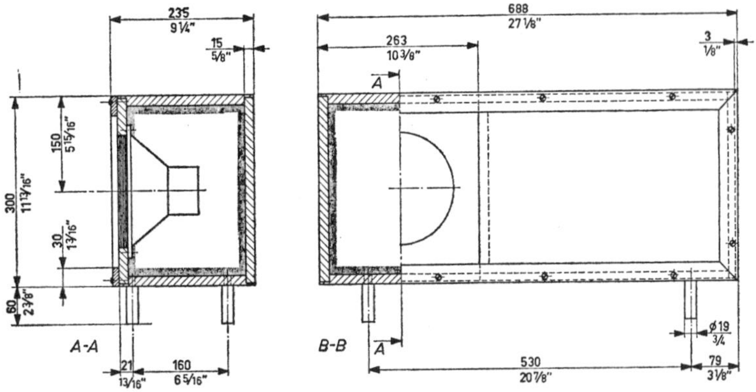


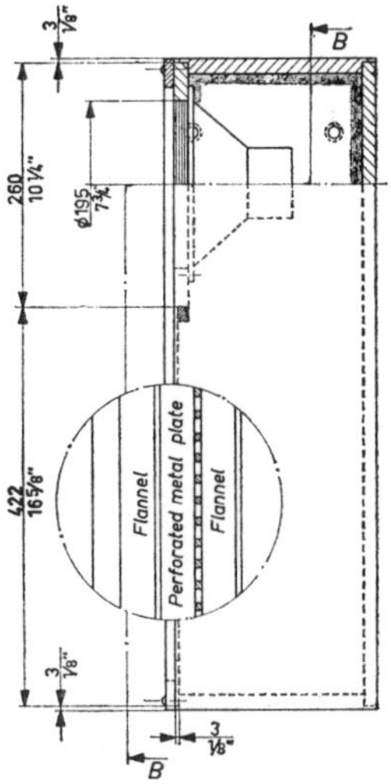


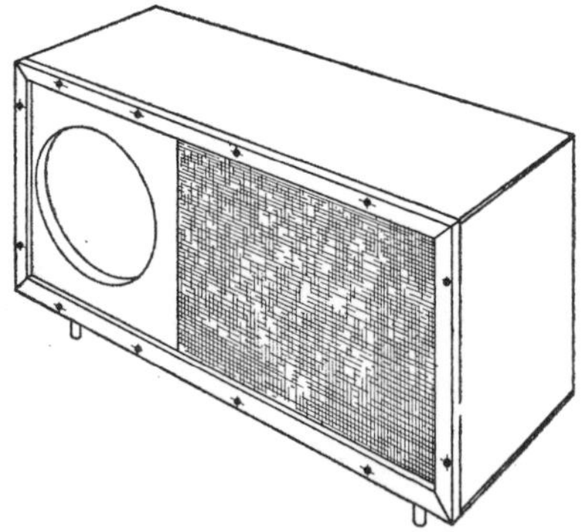

*Model 7*

Number of loudspeakers: 1

Speaker types: AD 3500/06, AD 3500/06 M or AD 3500/06 AM

*B, E*

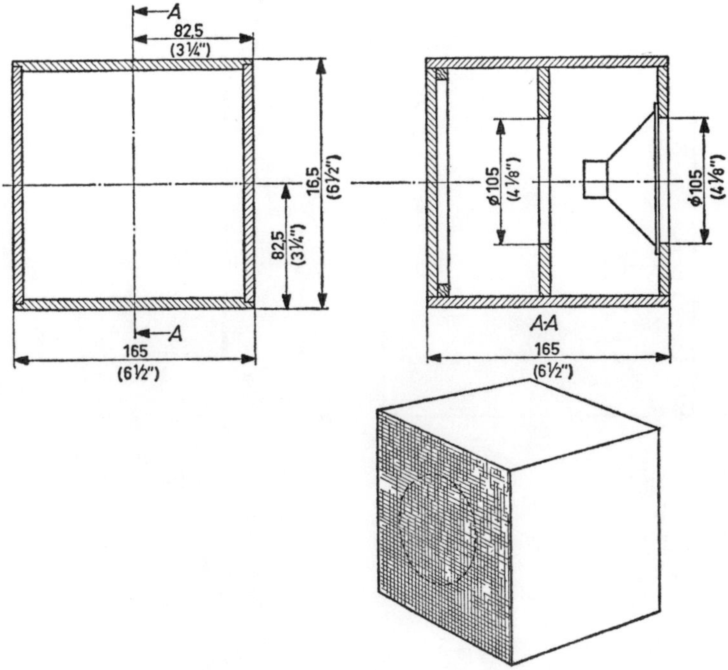

*Model 8*

Number of loudspeakers: 2

Speaker types: AD 3500/06, AD 3500/06 M or AD 3500/06 AM

*E*

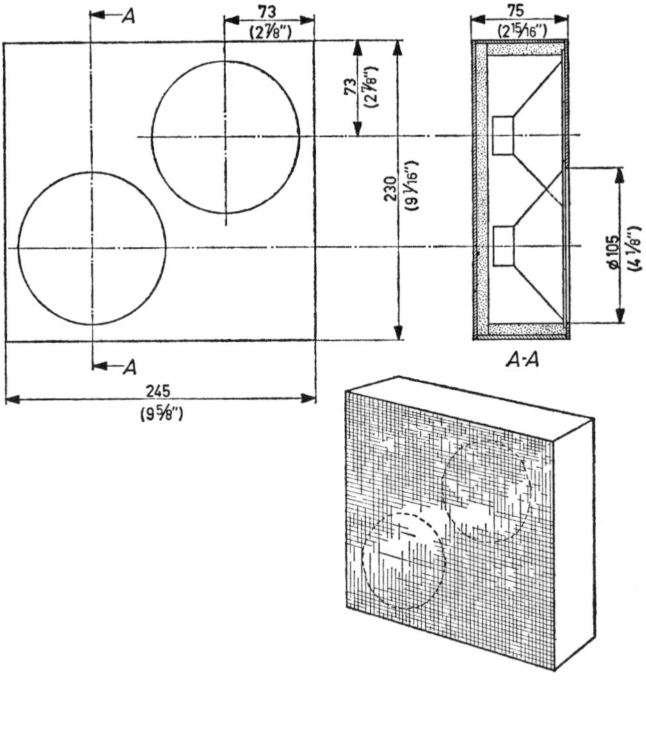

# *Model 9*

**Box suitable for all types of speakers by means of an interchangeable baffle.**

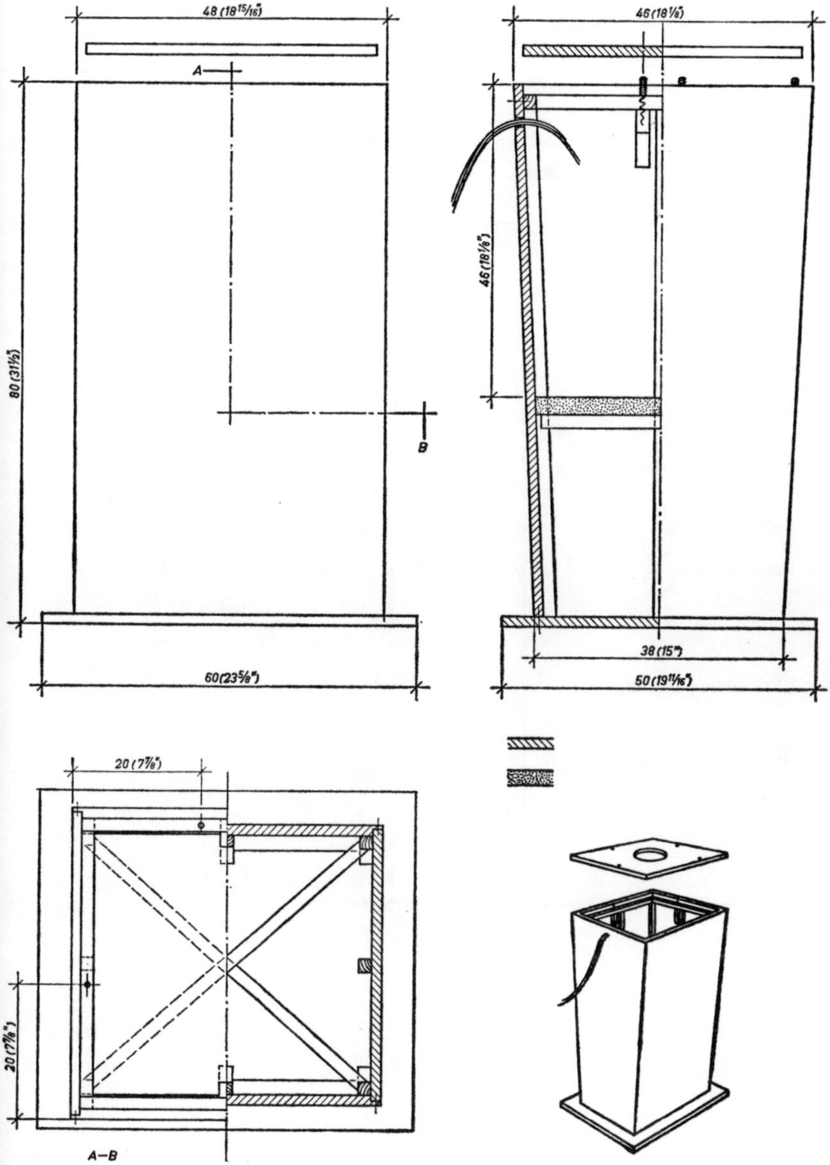

## APPENDIX

A survey of the values used in this book with the units on the MKSA scale in which they are expressed.

| *Value* | *Symbol* | *Unit* |
|---|---|---|
| voltage | $U$ | volts (V) |
| electric field intensity | $E$ | V/m |
| current | $I$ | amperes (A) |
| power | $P$ | (VA) = watt (W) |
| resistance | $R$ | (V/A) = ohm |
| capacitance | $C$ | (A·s/V) = farad (F) |
| inductance | $L$ | (V·s/A) = henry (L) |
| charge | $Q$ | (A·s) = coulomb (C) |
| magnetic flux | $Ø$ | (V·s) = weber (Wb) |
| magnetic inductance | $B$ | (V·s/$m^2$) = (Wb/$m^2$) = tesla |
| magnetic field intensity | $H$ | (A/m) |
| *BH* product | $BH$ | (Wb·A/$m^3$) |
| inductance constant | $\mu_0$ | (Vs/A·m) = (H/m) |
| relative permeability | $\mu_r$ | number |
| force | $F$ | newton (N) |

Table giving the dB values for several energy, current and voltage ratios

| $\frac{V_1}{V_2}$ or $\frac{I_1}{I_2}$ | $\frac{W_1}{W_2}$ | dB | $\frac{W_2}{W_1}$ | $\frac{V_2}{V_1}$ or $\frac{I_2}{I_1}$ |
|---|---|---|---|---|
| 1 | 1 | 0 | 1 | 1 |
| 0.891 | 0.794 | 1 | 1.26 | 1.12 |
| 0.794 | 0.631 | 2 | 1.58 | 1.26 |
| 0.703 | 0.501 | 3 | 1.99 | 1.41 |
| 0.631 | 0.398 | 4 | 2.51 | 1.58 |
| 0.562 | 0.316 | 5 | 3.16 | 1.78 |
| 0.501 | 0.251 | 6 | 3.98 | 1.99 |
| 0.447 | 0.200 | 7 | 5.01 | 2.24 |
| 0.398 | 0.158 | 8 | 6.31 | 2.51 |
| 0.355 | 0.126 | 9 | 7.94 | 2.82 |
| 0.316 | $10^{1}$ | 10 | 10 | 3.16 |
| 0.178 | 0.0316 | 15 | 31.6 | 5.62 |
| $10^{1}$ | $10^{-2}$ | 20 | $10^{2}$ | 10 |
| 0.0316 | $10^{-3}$ | 30 | $10^{3}$ | 31.6 |
| $10^{-2}$ | $10^{-4}$ | 40 | $10^{4}$ | $10^{2}$ |
| $10^{-3}$ | $10^{-6}$ | 60 | $10^{6}$ | $10^{3}$ |
| $10^{-4}$ | $10^{-8}$ | 80 | $10^{8}$ | $10^{4}$ |
| $10^{-5}$ | $10^{-10}$ | 100 | $10^{10}$ | $10^{5}$ |

Sound-pressure curves of the speakers described and recommended in this book.

AD 3500/06

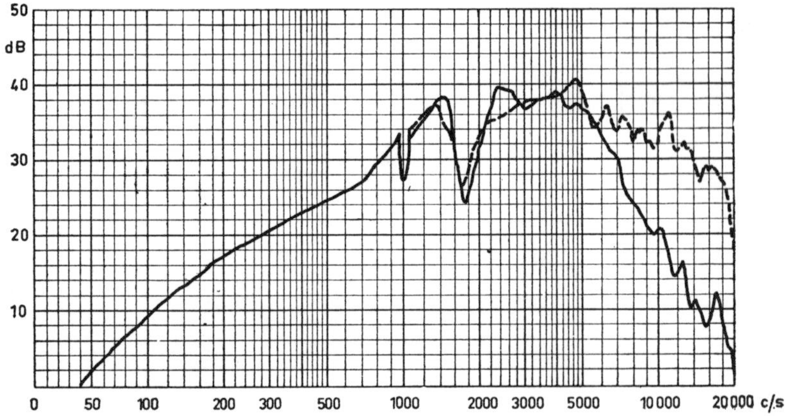


AD 3700/06

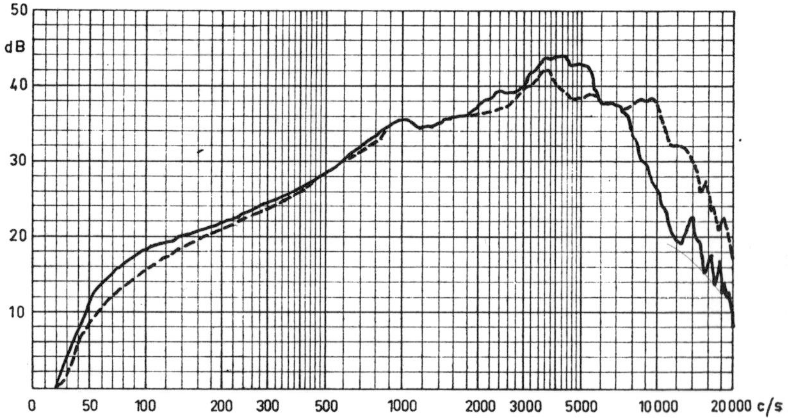

AD 3800/06

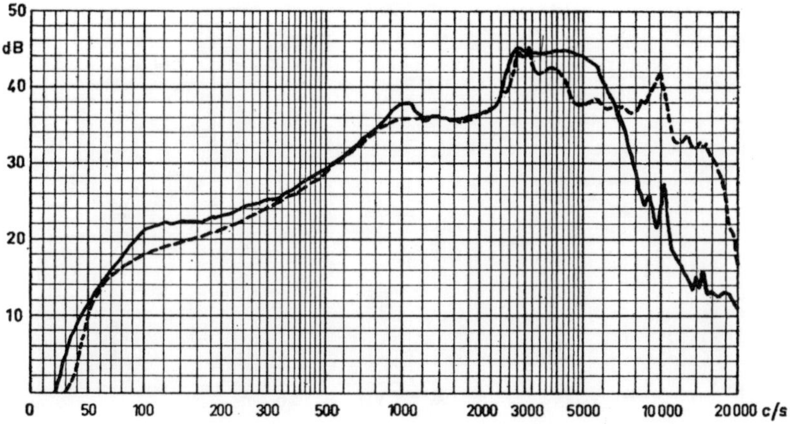


AD 4000

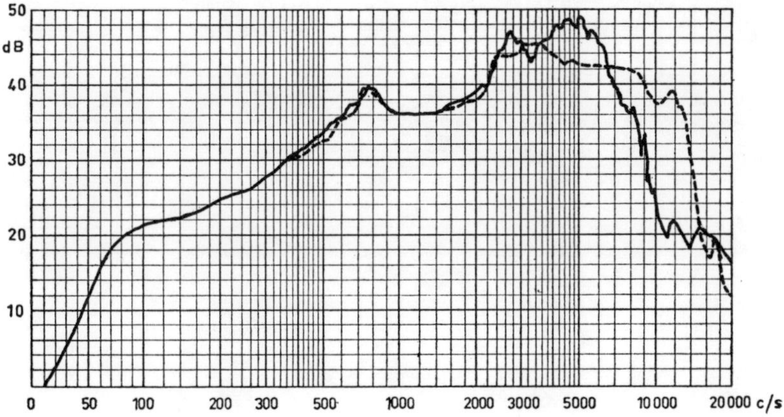

AD 5200

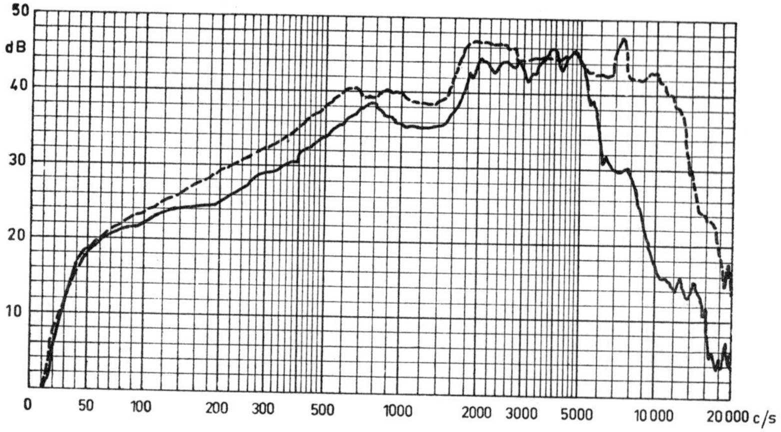


AD 4800

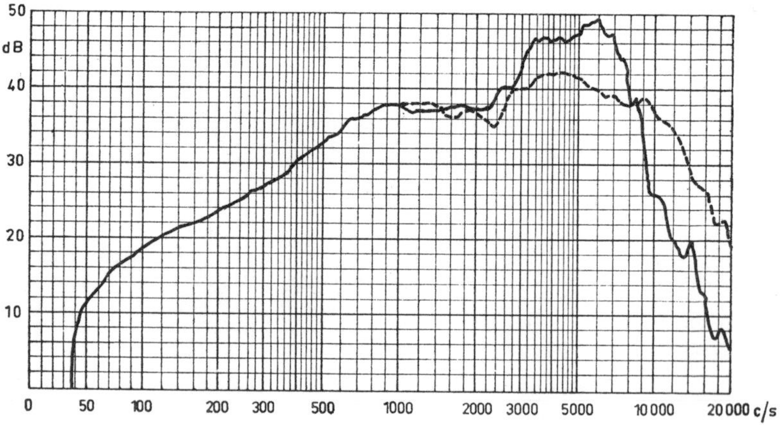

AD 4200

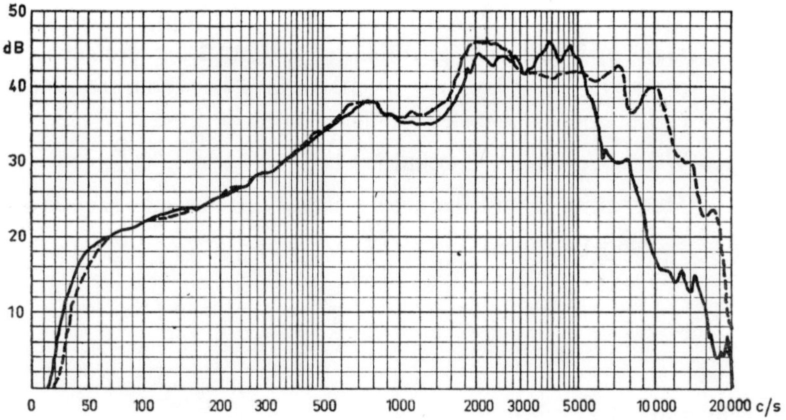

50
dB
40
30
20
10
0
50
100
200
300
500
1000
2000
3000
5000
10 000
20 000 c/s


9710

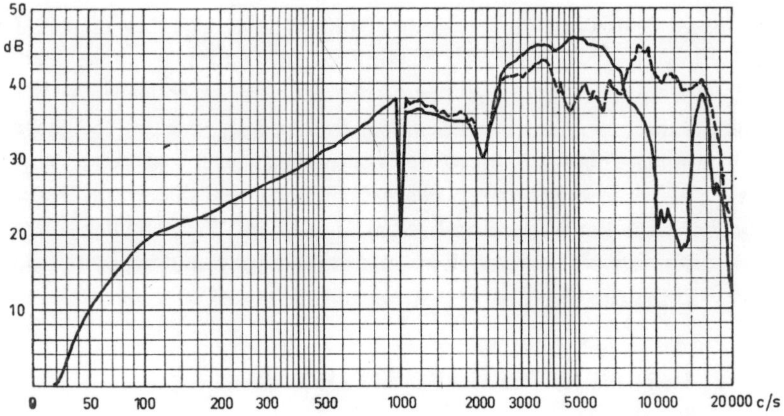

50
dB
40
30
20
10
0
50
100
200
300
500
1000
2000
3000
5000
10 000
20 000 c/s

## PHILIPS PAPERBACKS

- This is a series covering subjects in all branches of radio and electronics.
- Typical subjects covered are radio technology, audio fidelity, radio and television transmitting and receiving problems, tape recorders, DC relays, aerials, short wave reception, transistors in logical circuits, and electronic measuring devices.
- Written for both the technician and the student.
- Many titles to suit the keen amateur and the interested layman.
- Lavishly illustrated.

Already published in this series:

P 1 G. SLOT: **AUDIO QUALITY**

Requirements for high quality audio equipment.

VIII + 156 pages, 61 line drawings and 26 tables. 13s.6d. net

This book deals with the problem of obtaining the quality of sound reproduction that will give the maximum satisfaction to those listening to it. The author considers the whole problem objectively. A chapter spent on discussing the "problem" is followed by one on the technical specifications. This leads on to one on loudness level and power requirements. Then follow successively chapters on non-linear distortion, frequency response, pitch deviation, and background noises. Two chapters are then devoted respectively to stereophony and ambiophony, and finally the author deals with the classification of systems and the use of equipment.

P 2 C. G. NIJSEN: **THE TAPE RECORDER**

A complete handbook on magnetic recording.

X + 142 pages, 57 line drawings and 30 pages of plates, one a fold-out, on art paper. 13s.6d. net.

With this book, the reader wishing to make his own recordings can familiarise himself in a systematic way with both the theoretical and practical sides of this intriguing subject. At his disposal are the author's experience and extensive knowledge of the applications of the tape recorder, gained in professional sound studios and as an adjudicator at sound contests. Above all, this is a practical book on an aspect of electronics now in the forefront of public interest.

*Contents include:*

What is sound? - sound recording and reproduction - the tape recorder - acoustics - stereophony - choosing a recorder - advice on making recordings - applications of the tape recorder from A to Z - the recorder in general education and the study of music-tape clubs and sound recording as a hobby - dictation machines-magnetic recording conquers the world.

### P 3 D. J. W. SJOBBEMA: **AERIALS**

TV and FM receiving aerials.

VIII + 110 pages, 98 line drawings. 10s.6d. net.

Such things as high sensitivity and good-quality reproduction from a receiver deserve an aerial both correctly designed and erected. In addition to discussing the principles of aerial technology, the book deals fully with the many practical aspects of aerial arrays. The reader will find here not only many designs, but also solutions to those problems confronting every aerial erector.

*Contents include:*

The energy transfer from emitter to receiver - the receiving aerial - the choice and installation of the aerial - connecting the aerial to the receiver - attenuators - several receivers connected to the same aerial.

## RADIO SERVICING

Radio Servicing comprises a total of six books which together form a manual on the repair of radio receivers. All are written in simple language.

Each book deals with a specific subject and each chapter is followed by a short summary, together with a number of questions designed to test the reader's grasp of what he has read.
Theoretical principles are treated only where absolutely necessary. Clear insight is given into the functioning of a radio receiver, interpretation of technical diagrams, the use of measuring apparatus, tracing and repairing the most common faults, etc.

P 4 EDGAR J. BLACK: **DIRECT CURRENT AND MAGNETISM**

VIII + 119 pages, 92 line drawings. 10s.6d. net.

P 5 EDGAR J. BLACK: **ALTERNATING CURRENT AND ACOUSTICS**

VIII + 116 pages, 86 line drawings. 10s.6d. net.

P 6 EDGAR J. BLACK: **RADIO VALVES**

VIII + 126 pages, 90 line drawings. 10s.6d. net.

P 7 EDGAR J. BLACK: **A. F. AMPLIFICATION**

VIII + 109 pages, 82 line drawings. 10s.6d. net.

Other titles in Radio Servicing in preparation:
P 8 MEASURING INSTRUMENTS
P 9 FAULT-FINDING IN RADIO RECEIVERS

P 10 A. H. BRUINSMA: **CIRCUITS USING DIRECT CURRENT RELAYS**

IX + 86 pages, 66 line drawings. 13s.6d. net.

This book gives an insight into a number of uncommon circuits which have already proved themselves in practice and the possibility of applying direct current relays in commanded or programmed sequence circuits is fully explained.

*Contents include:*

General data of relays for direct current supply - operation of a relay - retarding a relay - short-term relay operation - operating a relay by means of physical phenomena - commanded sequence circuits - automatic sequence circuits with commanded start - oscillating circuits - some special circuits.

**P 11 J. Ph. KORTHALS ALTES: TRANSISTORS IN LOGICAL CIRCUITS**

VIII + 117 pages, 125 line drawings and 2 plates. 16s. net.

Starting from already known analogue relay circuits, this book describes in a special way the logical electronic circuits using semiconductors.

*Contents include:*

Some concepts of switch algebra - the twofold or binary system - the transistor as a switch - diodes - AND-, OR and NOT operation with semiconductors - signal amplification - bistable stages - multivibrators - counting switching circuits - arithmetical operations - observations - examples of applications - practical hints.

**P 12 P. W. van der WAL: LOUDSPEAKERS AND LOUDSPEAKER CABINETS**

VIII + 110 pages, 156 line drawings and 3 plates. 15s. net.

Besides providing particulars of numerous different speaker boxes, this book is a source of information on various matters which, although apparently incidental, must nevertheless be given due consideration if the best results are to be obtained. To enable the "do-it-yourself" enthusiast to deal with certain problems which may arise during the construction of the boxes, a chapter of practical hints is also included.

**P 13 J. VASTENHOUD: SHORT WAVE LISTENING**

VII + 120 pages, 33 line drawings and 4 plates. 12s. 6d. net.

The book, which deals with the possibilities and problems of short-wave reception on the level of popular science will enable the reader to discover a whole new world of his own.

*Contents include:*

Short waves - the principles of short wave transmission - practical short wave transmitting - short wave prediction - sources of interference - the aerial - the correct choice of receiver - communications receivers - do any regulations exist governing the use of frequencies in the short wave band - DX-ing in practice - DX-ing with a tape recorder - DX-ing using a frequency meter.

Fig. 9. Photograph of a vibrating speaker. The outer portion of the cone is "at rest"; the frequency fed to the speaker is about 3500 c/s. (See p. 5).

Fig. 22. The "dead" room (see p. 17).

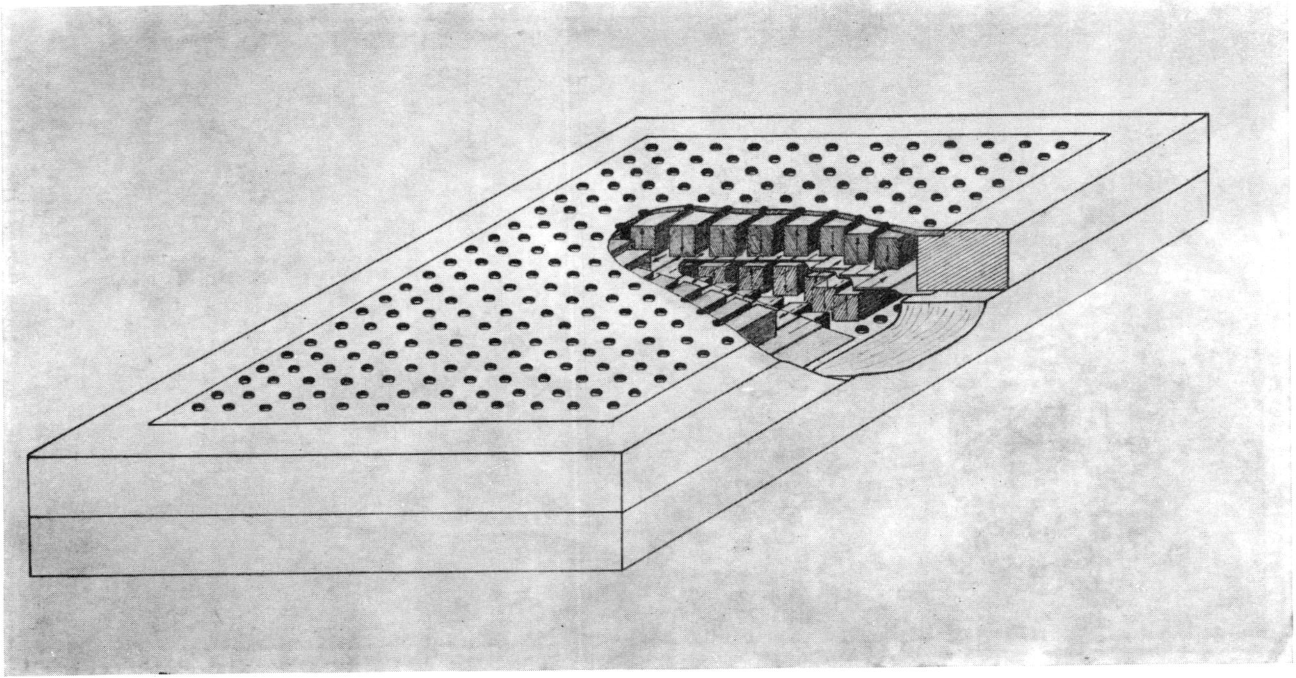

Fig. 34. Isophase speaker (see p. 21).